EVOLUTION OF THE GAS INDUSTRY

Lambeth Gas Works, London, 1872 (*wood engraving by Gustave Doré*)

EVOLUTION OF THE GAS INDUSTRY

Malcolm W. H. Peebles

New York University Press. New York *and* London

First published in the U.S.A. by
New York University Press,
Washington Square, New York, N.Y., 10003

Printed in Hong Kong

Library of Congress Cataloging in Publication Data

Peebles, Malcolm W H
Evolution of the gas industry.

Includes bibliographical references and index.
1. Gas industry — History. 2. Gas, Natural.
I. Title.
HD9581.A2P43 338.2′7285′09 80-19311
ISBN 0-8147-6580-7

Contents

List of Tables

Preface

In the beginning man relied solely upon muscle power, that of his own and no doubt that of animals as well. Then, at some unknown point in time, probably many thousands of years ago, prehistoric man discovered how to make fire and to burn wood and other primitive fuels such as dried peat and animal dung to keep him warm and to cook his food. The first use of coal is not recorded, but no doubt this also dates back several thousand years—certainly it was in relatively common use by Roman times. However, it was not until the Industrial Revolution that energy began to be used by man on a massive and widespread scale.

The Industrial Revolution was fuelled principally by coal. Thereafter coal remained king of the energy scene until shortly after the Second World War when it was displaced by oil, which then became the world's largest individual source of primary energy supply. Today oil supplies about 44 per cent of the world's energy requirements, or 55 per cent if the communist countries—which still rely heavily on coal—are excluded.

Natural gas, with a few exceptions, notably in the United States, is a relative newcomer to the energy scene compared with oil and in its turn has largely displaced manufactured gas (by definition a secondary form of energy) during the last two to three decades. One illustration of the rapid growth that natural gas has achieved over this period is Western Europe. In 1965 the contribution made by natural gas to total primary energy consumption was barely 2 per cent, but by 1977 this had expanded to over 17 per cent.

This remarkable increase in the use of natural gas in Europe, mirrored to a lesser or greater extent in a number of countries elsewhere in the world, has been realised primarily at the expense of coal. For example, in Western Europe coal's share of primary energy consumption for the same two years of 1965 and 1977 declined from 47 per cent to 24 per cent, in spite of the overall increase in energy consumption over this period. This was not just a decline in coal's percentage contribution, but a substantial reduction as well in the absolute volumes of coal consumed. Moreover, apart from the displacement of coal *per se*, natural gas in replacing manufactured gas made largely from coal represented the loss of a further important outlet for coal as a gas-making feedstock.

Although coal was probably the principal sufferer from the inroads of natural gas, the expansion in the use of natural gas has also been partly at the expense of the growth rates that oil products, in particular fuel oil and gas oil (and also electricity), might otherwise have enjoyed but for the introduction of natural gas. Admittedly, various factors have favoured natural gas in many countries, including its low price relative to the prices of competing fuels, and its cleanliness and hence desirability as a fuel in a world growing ever more conscious of its environment. Furthermore, it should not be overlooked that indigenous natural gas, where available, was often preferred and received some form of encouragement over imported oil for political, security of supply, and balance of payment reasons. This was not true everywhere, but it was certainly the case in a number of countries.

The size of the world's proven natural gas reserves, coupled with high expectations for substantial new discoveries, indicate that supply availability could continue growing well into the next century. On the other hand oil supplies, based on existing reserves and future expectations, could well plateau out at an earlier date. Newer forms of energy such as nuclear power, solar, wind or tidal energy, have either yet to make a significant contribution, or pose considerable practical problems for which solutions will not be found overnight.

All this points to a bright future for a natural-gas-based gas industry. At the same time a return to a gas industry relying, as it did until fairly recently, on complicated and expensive manufacturing processes for converting solid and liquid fuels to combustible gases seems remote, other than the use of high calorific value manufactured gas as a supplement, where necessary and appropriate, to natural gas supplies.

The balance of probability is, therefore, that the world's gas industry will for many years to come rely increasingly or exclusively on natural gas as its prime source of supply. This book attempts to record, before memories fade and the facts become too obscured or lost with the passage of time, how the gas industry has evolved over the years since its creation in the early nineteenth century. Attention has been concentrated on certain selected countries. In each selected case the gas industry has developed in a unique way conditioned by local circumstances; no one country or market has followed a pattern of development which may be regarded as being typical for the industry as a whole. Obviously, there are other countries not described in this book where the pattern of development has not differed too dramatically from one or other of the countries reviewed, but even so there are usually subtle differences of consequence which the serious student would be well advised to investigate in depth before arriving at any premature conclusions.

Each and every gas market has its own peculiarities of supply, government involvement, or lack of involvement as the case may be, pricing, and competition with other energy forms, exacerbated by the social and environmental conditions applicable therein. The gas business may be becoming more international in character, but it still falls a long way short of some of the more general influences that govern the conduct of the international oil and coal businesses which have received so much publicity and informed opinion in recent years.

Little attempt has been made to look too far into the future as we are concerned here primarily with what has happened in the past. I hope this book will serve as a factual base from which some readers may wish to start considering how the future conduct and prosperity of the business may evolve.

Claygate, Surrey M.W.H.P

Acknowledgements

I wish to express my gratitude to many friends and colleagues in the gas industry around the world whose interest, encouragement and assistance has helped to make this book possible. I am particularly indebted to Lou Brocker, Tony Budd, Charles Case, Yuri Domrachev, Drew Hamilton, Minoru Ikushima, Alan Levine, Brian McLellan, Jeremy Russell, Eric Short, Don Stephenson, and Jan Visser, either for their opinions and constructive comments, and/or for the trouble they have taken to seek out suitable reference material and illustrations for me. My special thanks are also due to Anne Luget for her excellent and patient typing of my manuscript.

In addition, a number of national and international organisations have kindly made available to me various books, reports, illustrations, and the like, and, where appropriate, have granted me permission to use certain copyright material. Their ready co-operation in these matters has been of considerable assistance; in particular my thanks to:

Atlanta Gas Light Company
British Gas Corporation
Brooklyn Union Gas Company
Conch Methane Services Limited
Gas Shiryo-kan
Gotaas-Larsen Inc.
Japan Gas Association
N.V. Nederlandse Gasunie
Oil and Gas Journal
Osaka Gas Company Limited
Petroleum Economist
Ruhrgas Aktiengesellschaft
Shell International Gas Limited

Southern Natural Gas Company
Sojuzgazexport
Tokyo Gas Company Limited

Finally, I wish to make it clear that the responsibility for any errors, significant omissions, expressions of opinion, interpretations of facts and statistics, rests entirely with me. This book has been a personal project and as such does not purport to represent the views of any particular organisation.

Introduction

In most industrialised countries low calorific value gas manufactured principally from coal feedstocks, and distributed generally at low pressures for public consumption and to small industrial consumers for fuel purposes through localised pipeline grids, is either being or has already been displaced by high calorific value, high-pressure natural gas. Undoubtedly the availability of natural gas, with its inherent economic, qualitative, technical and other advantages over manufactured gas, has been the saviour of the gas industry in recent years. It is doubtful whether the traditional manufactured gas industry could have survived, in the face of increasingly severe competition from liquid petroleum fuels and electricity, if it had had to rely upon manufactured gas indefinitely.

On the other hand manufactured gas has paved the way for the rapid introduction of natural gas in the post-Second World War years—earlier in the case of the United States where natural gas was developed almost simultaneously with manufactured gas. But for the existence of the distribution and marketing infrastructure built up painstakingly in past years for manufactured gas, natural gas would have had an appreciably harder task in establishing the position it now enjoys, or in the newer markets, what it can expect to achieve in the coming decades.

This book, as the title implies, endeavours to trace and recount the evolution of the gas industry from its birth in the early nineteenth century through to the latest date for which statistical and other factual data are currently available (variously 1976 to 1978). Certain countries have been selected for detailed examination as it is not practicable to attempt to cover every country and facet of the business in one book. The countries selected are among the more important markets for gas, it being appreciated that there are a number of other significant gas markets in the world today which unfortunately have had to be omitted on this occasion.

The United Kingdom deserves selection as it was in England that the world's first commercial gas enterprise was founded in 1812. Apart from being the pioneer of various new developments and processes, the UK has also gone through the transition from manufactured gas to natural gas, and from an essentially privately owned industry to a business which, apart from gas production, is now entirely owned and controlled by the state.

The United States is an obvious choice as this is the world's largest gas market. Uniquely both manufactured gas and natural gas were developed in this country almost simultaneously along parallel but separate lines until eventually manufactured gas was almost totally supplanted by natural gas. Although the gas business in America is a private industry affair, it is nevertheless subject to one of the highest levels of federal and state regulation experienced by almost any type of private industry in the western world today.

Japan, yet again, is quite different. One of the world's most rapidly growing gas markets comprising both manufactured gas and imported liquefied natural gas. Japan can also claim to have both the smallest and largest privately owned gas enterprises within its boundaries. The increasing reliance on imported gas is a further feature of the Japanese scene.

The Netherlands provides another contrast. The famous Groningen field is currently the world's largest gas field under full-scale commercial development. Dutch gas, including exports, is at present the largest single supply source of natural gas in Western Europe, and as such has had a profound effect on the gas industries in Belgium, France and West Germany. Interestingly, the gas business in the Netherlands is also conducted largely by a combination of private industry and government interests, having previously been variously a private industry and a wholly owned state activity.

The USSR is the world's second largest gas market and cannot, therefore, be overlooked. For completeness sake, a brief discussion is also included in this particular chapter of various European gas markets which import Soviet gas.

The final chapter is devoted to one of the more recent and exciting gas industry developments, namely the evolvement of liquefied natural gas, both as a means of supplementing local gas supplies, and as a major international source of gas supply in its own right.

Before embarking upon these country reviews, a generalised description is given of the historical origins of the gas business. This discusses some of the more noteworthy earlier discoveries and experimentations made prior to the establishment of the gas industry proper in the early nineteenth century.

There are several difficulties in writing a book of this nature. First and foremost perhaps is the extent to which one can rely on published material relating back to the last hundred years and more. There is the ever-present danger in such circumstances of recycling apparent facts from more recent publications, but which were in reality perhaps merely opinions or best guesses of what happened rather than true factual accounts. The author has endeavoured to sift out the former from the latter to the best of his ability, but does not claim to be infallible on every count. Where discrepancies exist in published literature, every attempt has been made to select what appears to be the most authenticated and reliable account.

The second problem is that of what units of measurement, currency, etc., to use. In this regard the temptation to convert all data to a common basis has been resisted for two reasons. First, for some of the early data the underlying facts, for example as to heat content (calorific value) or volumetric conditions of measurement, are not known. Second, there have been so many fluctuations in currency exchange rates over the years that to convert all financial data to one common currency unit would in fact distort the position when viewed in present-day terms. Accordingly the author has quoted all data in their original form leaving it to the reader to convert and interpolate, if he so wishes, as he thinks fit.

The third difficulty is that of terminology and abbreviations, some of which are unique to the gas industry and as such are not always readily comprehensible to those less familiar with the international gas business. Wherever possible the text describes briefly such gas expressions, and abbreviations are usually spelt out in full in the first instance. However, in order to aid comprehension, a glossary of some common gas industry terms, and four tables of natural gas equivalents to facilitate conversion of one unit to another, have been included at the end of this book.

A final point: the author has used, where appropriate, the expression 'milliard' to denote a thousand million (i.e. ten to the power of nine), rather than 'billion' which is used more frequently in American literature. Milliard has the merit of not being ambiguous, whereas in some countries billion can be taken to mean either a thousand million or a million million. Perversely, however, the expression 'trillion' has been used in the specific context of the United States where it is always taken to mean a million milllion (i.e. 10 to the power of 12). The unambiguous and precise units as recommended by the International Gas Union are now being adopted and introduced in a number of countries, but it will be some years yet before they are applied everywhere. Accordingly, and as explained above, the author decided to use the original units given in references of the literature.

1: The Historical Origins of the Gas Industry

> And the angel of the Lord appeared unto him [Moses] in a flame of fire out of the midst of a bush: and he looked, and behold, the bush burned with fire, and the bush was not consumed.
>
> Exodus 4:2

Ancient Times

No history of the evolution of the gas industry would be complete without at least passing mention of some of the references in folklore, mythology, religion and literature to the existence of such things as will-o'-the-wisp, burning bushes, perpetual lights, and burning springs in ancient times. While the nature and composition of the gases that gave cause to these phenomena were not known in those days, there can be little dispute that the gases in question occurred naturally and were predominantly methane—in most instances they were probably seepages of natural gas accidentally ignited by lightning or by man.

One of the most quoted references is of the use of (natural) gas by the Chinese possibly three thousand years or more ago for the heating of pans of brine water to obtain salt. Both the Romans and the Greeks, among others, are said to have known of the existence of such gases before the birth of Christ. But there is no evidence to suggest that, apart from the Chinese, they put such seepages of gas to any practical use beyond that of channelling it in some way or another to create ever-burning sacred lights or flames for religious purposes.

Plutarch (A.D. 60–140) mentioned a lamp which had remained alight for centuries. St Augustine (A.D. 354–430) described a perpetual lamp in the Temple of Isis which neither wind nor water could extinguish. In 1550, in a vault on the island of Nesis, near Naples, there was found a still-burning lamp which had been there from the beginning of the Christian era. And in the Middle Ages a perpetual lamp was found in the tomb of the father of Constantine the Great in Yorkshire, where it had been burning since the third century. Many sightings of burning springs by such famous figures as Julius Caesar and George Washington are recorded in literature.

1.1 The Chinese were reputed to have used bamboo pipes some 3000 years ago for transporting natural gas. This photograph from more recent times shows the use of bamboo pipes in the province of Szechwan

Fascinating though these events and occurrences may be, they in fact have little direct relevance to the birth and subsequent development of the gas industry as we know it today. However, before embarking on an account of these matters, it is appropriate to record the origin of the word gas.

Gas Gets Its Name

The word 'gas' was invented by the Flemish scientist Jan Baptista van Helmont (1577–1644). In his work *The Origins of Medicine* he presented the findings of his experiments (*c*. 1609) during which he had discovered that a 'wild spirit' escaped from heated coal and wood. He concluded that something had escaped and stated: '*hunc spiritum, incognitum hactenus, novo nomine gas, voco*', or, 'to this vapour, hitherto unknown, I give the new name "gas".' The derivation of the name is generally thought to be the Greek word 'chaos' as van Helmont also wrote: '*paradoxi licentia, in nominis egestate, halitum illum, gas vocavi, non longe a chao veterum scretum*', or, 'in the absence of a name and contrary to normal practice, I have given to this vapour, which differs little from "chaos of the ancients", the name of "gas".'

Apart from discovering that materials lost weight after being heated, and distinguishing gases from solids and liquids, van Helmont's further claim to fame was that he was the first to identify carbon dioxide as a separate substance.

In this account we shall be concerned mainly with manufactured gas—combustible gases derived from primary energy sources (e.g. coal or oil) by processes involving chemical reaction, and natural gas—a predominantly hydrocarbon gas of natural origin found in underground structures similar to those containing crude oil. Town gas, a more generic expression, historically usually meant manufactured gas as piped to consumers, but it can also refer to the distribution of natural gas or mixtures of manufactured and natural gases as well: the term is less precise in its meaning than manufactured gas or natural gas and is now gradually falling into disuse.

The Discovery of Coal Gas

It is not possible to pinpoint the discovery of manufactured gas with precision. Literature records many reported experiments in Britain, France, Belgium, Germany and elsewhere in the seventeenth century which resulted in the discovery of combustible forms of gas made from coal, wood and peat. For example, it is recorded that around 1681 Professor Johann Becker of Munich, Germany, discovered that combustible gases resulted from the heating of coal in the absence of air. About 1684 the Reverend John Clayton (1657–1725) of Wigan, England, collected gas he had made from heating coal and amused his friends by lighting the gas as it escaped; he called this gas 'Spirit of Coals'.

Becker and Clayton are but two prominent persons among others who have been credited with the invention of combustible coal gas in the seventeenth century. It is more than probable that several researchers working quite independently and unbeknown to each other made much the same sort of discovery at around the same time.

Early Pioneers

There is less controversy concerning the role played by the Scottish engineer, William Murdock (1754–1839) of steam-engine fame, who is regarded by many as 'the father of the gas industry'. Murdock (also known as Murdoch) was born at Bello Mill, near Muirkirk, a remote village some 30 miles south of Glasgow. He first produced gas by heating coal in a teapot borrowed from his mother. However, unlike most of his predecessors, he innovated various practical developments in making, purifying and storing gas from coal during his subsequent employment with Boulton and Watt. Murdock is reported to have illuminated his house in Redruth, Cornwall, with gas in 1792, and five years later, to have installed a gas light outside the door of the Manchester Police Commissioners. This was followed by illuminating the exterior of the Boulton & Watt factory in Birmingham, England, in 1802 and a large cotton mill in Salford, Lancashire, in 1805. Records indicate that the latter cost £600 a year compared with £2 000 per year for candles.

Muirkirk, apart from being the birthplace of Murdock, also has another more recent minor claim to fame. Quite accidentally Muirkirk was the last place in Britain to be converted by the British Gas Corporation to natural gas. This was on 19 September 1978, when the local gasworks built in 1857 was finally shut down and its 570 customers switched to natural gas, with one 80-year-old customer at last having to extinguish the manufactured gas lights which had lit his family's home for over 100 years. The end of an era—at least in Britain—but that is another story to be told in a later chapter.

Reverting, however, to this account of the early pioneers, around the time that Murdock was undertaking these developments, others elsewhere were carrying out demonstrations and experimenting with gases made from coal and other materials. Professor Jean Pierre Minkelers, a Dutchman, lit his lecture room at the University of Louvain in 1783 with coal gas. In 1779, George Dixon used coal gas to illuminate a room of his house, and Lord Dundonald lit up Culross Abbey with coal gas in 1787. In France, Phillipe Lebon (1767–1804) produced gas from sawdust, patented a gas fire on 21 September 1799, and held public demonstrations of gas lighting in 1801, but it was not until 1819 that gas lights appeared in the streets of Paris.

In the United States, Benjamin Henfrey demonstrated the use of gas for lighting in Baltimore in 1802 and again in the following year in Richmond, Virginia. There were other demonstrations, but David Melville of Newport, Rhode Island, seems to have been the first American to have gone beyond the demonstration stage when he lit his own house in 1807 and illuminated several cotton mills between 1813 and 1817.

Obviously there were many people in a number of countries experimenting in various ways with the use of combustible gases derived mainly from coal around the turn of the nineteenth century. However, such evidence as exists tends to point to the fact that Murdock was perhaps the leading innovator at

1.2 William Murdock (1754–1839) using gas lights at his home in Redruth, Cornwall, England in 1792. From a Science Museum diorama (*British Gas Corporation*)

this time in taking these early experiments and demonstrations a stage nearer to commercial practicality.

The First Commercial Enterprises

The first commercial use of gas is difficult to determine, if only because of the interpretation that can be put on the meaning of commercial. A UK government Committee of Enquiry records that the first gas company—the London and Westminster Gas Light and Coke Company—came into being on 30 April 1812. Another authoritative source states that this company built the world's first gas works in Great Peter Street, Westminster, London, in 1812, laid wooden pipes and illuminated Westminster Bridge with gas lamps on New Year's Eve in 1813. The establishment of this company is now generally regarded as being the start point of the world's gas industry in a commercial or business sense.

However, no historical account of the gas industry would be complete without mention of Friedrich Albrecht Winzer (1763–1830) who was born in Moravia. He came to London in 1803, changed his name to Frederick Winsor, and patented a gas lighting system. He installed gas lights outside Carlton House, Pall Mall, the Prince of Wales's London home, in June 1807. It was Winsor in fact who founded the London and Westminster Gas Light and Coke Company. Winsor was not a very efficient businessman and it was not long before Samuel Clegg (1781–1861) took over the running of the company from him. Among other innovations, Clegg designed the horizontal rotary retort in 1817.

The United States was not far behind in these developments. On 17 June 1816, Rembrandt Peale and four other prominent local men, formed the Gas Light Company of Baltimore, the first manufactured-gas company in America. The company was formally incorporated on 5 February 1817, and the first street light was supplied with gas two days later. This was followed by the Boston Gas Light Company in 1822, and the New York Gas Light Company in 1825. Incidentally, the gas supplied initially by the Gas Light Company of Baltimore was produced from the distillation of pine tar. This process was employed until 1822 when the company decided to adopt the standard English practice of manufacturing gas from coal, with much more satisfactory results.

In Germany, Franz Dinnendahl, a steam-engine manufacturer, illuminated his factory at Steele, near Essen, with gas in 1818, and the first German gas works was built in Hannover in 1825. By 1870, 340 German gas works were producing gas from coal, wood, peat and other materials for public lighting, domestic and industrial uses.

From these and other accounts it is clear that the commercial—in this context taken to mean sales of gas to members of the public and to industrial enterprises—use of gas began in the early part of the nineteenth century, about 120 years or so after coal gas was first discovered. It is also evident that the United Kingdom probably took pride of place in these early days of development, with the United States and Germany hot on its heels.

Uniquely, as the manufactured gas industry in America was getting underway, a separate but parallel development of natural gas was taking place almost simultaneously. The first recorded commercial use of natural gas was in Fredonia, New York, in 1821 only four years after the Gas Light Company of Baltimore commenced operations. This development is discussed in greater detail in Chapter 3. Natural gas was also used to a very limited and localised extent in Canada in 1861, and in one or two other countries around this time, but these developments were not to become of any significance until many years later.

Reverting to the genesis of the manufactured gas industry, by 1819 Brussels and other Belgian towns had gas lighting. This was followed by the introduction of manufactured gas in many other cities and countries: Sydney, Australia 1841; Switzerland 1843; Sweden 1846; Toronto, Canada 1847; in the 1850s, Denmark, Spain, Italy, Poland, Malta, Rio de Janeiro, Buenos Aires, Calcutta, Mauritius, Hawaii, and parts of Mexico; Adelaide 1863; Austria, Hong Kong and Singapore in 1864, and in Japan in 1872—the list is not complete, but illustrative of the rapid spread of gas in the second half of the nineteenth century.

The Incandescent Mantle and New Applications Introduced

The general introduction of the electric dynamo in the 1870s represented a real threat to the gas industry as the principal application for gas at that time was illumination. In the early years gas was burnt in open-flame burners. This was inefficient and gave an indifferent, varying quality of light. However, at about this time Dr Carl Auer von Welsbach of Vienna invented the incandescent gas light mantle. Welsbach established the first factory to make gas mantles in London in 1877, but it was not until 1895 that the quality of mantle was such that they could be used for street lights. Some years before this—in 1826—the world's first gas cooker was devised in England by James Sharp, but it was not until 1851 that such equipment came into use in America. Water heaters, room heaters and many other ancillary appliances such as soldering irons and hair-curling tongs appeared on the scene in the mid to late 1800s.

A very important invention was made in 1855 by the German scientist, Robert Wilhelm von Bunsen (1811–99), based on earlier work by Faraday and other scientists. It was the atmospheric burner in which a jet of gas aspirated part of the air for combustion into a mixture before reaching the burner ports and flame zone. This significant invention opened the way for gas to be used as a fuel for a whole variety of industrial and other applications. The original Bunsen burner has, of course, been modified and improved, but its basic concept has changed little over the years and remains in wide use in laboratories, schools and such establishments to this day.

1.3 Gas works, Cologne, West Germany erected by the Imperial Continental Gas Association, 1841

1.4 An old-fashioned gas boiler of the nineteenth century (*British Gas Corporation*)

Improvements in Manufacturing Processes

Development and inventive effort were not, of course, confined to end-use applications; processes for the manufacturing of gas received equal attention. One new process, which was destined ultimately to outstrip the production of gas by the distillation of coal in America, but not necessarily in other countries, was developed in the early 1870s by Professor Thaddeus S. C. Lowe of Norristown, Pennsylvania. This process produced so-called water gas. Anthracite or coke were placed in a furnace or generator and brought to incandescence by being blown with air; products of combustion were vented.

1.5 A British 'Black Beauty' gas cooking stove by R. & A. Main, *c.* 1878 (*British Gas Corporation/Science Museum*)

After some minutes the blast of air was cut off, and steam under pressure was blown into the bed of glowing fuel. This decomposed the steam into hydrogen and oxygen; the latter combined with the incandescent carbon to form carbon monoxide. The resultant gas, comprising hydrogen, carbon monoxide and certain other substances, had a calorific value of about 300 Btu per cubic foot, but very little illuminating power because it burned with a blue flame. In order to increase its calorific value and illuminating qualities, the gas was fed into a carburettor, a brick checker work chamber. Oil was injected, vaporised and mixed with the gas. This enriched gas was called carburetted water gas. Purification, storage, and so on, were the same as for gas produced by coal carbonisation. Lowe was granted a patent in 1873 and the first plant was installed in Phoenixville, Pennsylvania, in 1874. Its outstanding merit was that within acceptable limits the calorific and illumination values could be regulated to provide a gas of consistent quality.

Prior to Lowe's development, considerable improvements had been made on the original batch-filled horizontal iron retorts in which coal was carbonised. These evolved into 'through retorts' introduced by another Lowe, George Lowe, in 1831, in which the coke produced could be withdrawn at the end opposite to that at which the coal was fed. In 1868, automatic stoking machines were introduced. And in 1885, André Coze of Rheims devised the inclined retort enabling charging and discharging of the retort to be carried out under gravity.

A major development was the introduction of the vertical retort. The great advantage here was continuous operation with coal being fed in at the top and coke withdrawn at the bottom. Thaddeus Lowe's process described above to produce water gas and carburetted water gas was a further refinement of these processes.

Pipes, Metering and Storage

An important contribution from Germany was the development of the Mannesmann process for the manufacture of seamless steel pipes. These pipes were suitable for high, by the standards of that time, internal gas pressures and replaced over time pipes that were made of copper, lead, cast iron and wrought iron; these in turn having replaced the original wooden pipes in which gas was first distributed.

Gas metering also received attention. The first known patent for a wet gas meter was by Samuel Clegg in 1815, but records indicate that the metering of gas did not come into general application in England or America until the 1830s. The original meters had a rotating drum operating in a bath of water and these caused problems, either in areas where the water could freeze during the colder months of the year, or because the water evaporated. But by around 1850, dry meters devised by John Malam (1820), by Miles Berry (1833) and by Alexander Wright (1844) were introduced and these remain the basis of the type of meter in general use today. The introduction of prepayment meters in the United Kingdom, patented by T. S. Lacey in 1870, not long after was an

1.6 Low-pressure gasometer for manufactured gas as used in West Germany before 1930 (*Ruhrgas A. G.*)

interesting development in that it helped to spread the use of gas to the poorer sections of the community

In the early days gas holders were not in general use. Many gas works preferred to discharge any excess gas to atmosphere during periods of low offtake, i.e. mainly in daylight hours as gas was used largely for illumination. Within a few years gas holders were becoming a necessity for economic reasons. Between about 1816 and the turn of the century a variety of gas holders were devised—again Samuel Clegg was in the forefront of such innovators—almost all of which required water to act as a seal between the separate sections of the holder. This was a problem as it meant the gas had to be dried before distribution and there was the further complication of disposal

1.7 A wagon for delivering meters to customers' premises, Atlanta, Georgia, *c.* 1900 (*Atlanta Gas Light Company*)

of the contaminated water. It was many years later before waterless holders were introduced.

It is interesting to record that the original gas holders were graduated so that the volume of gas contained therein could be calculated. These holders became known as gasometers, a name which has survived to this day.

The Pattern of Development Post-1900

The above brief description of the historical origins of the gas industry takes us up roughly to the early 1900s. The industry did not, of course, remain static thereafter, but the pattern of development varied increasingly from one country to another and thus does not lend itself readily to a generalised account. In most countries, gas for illumination was displaced steadily by electricity. However, it is not entirely eliminated to this day, quite apart that is from the more recent development of natural gas lighting for decorative and certain other specialised outdoor uses. Street gas lights were in service in some parts of London and Birmingham as recently as 1975 and still remain in service in various other countries including the UK.

As the illumination market declined so this was more than replaced by the increased usage of gas for such domestic purposes as cooking, water heating and space heating. The industrial use of gas also grew, but in no way did the growth match that of coal up to the late 1930s, or that of oil thereafter. We are still speaking here of manufactured gas as the growth of natural gas in North America over this period was quite a different story. Generally speaking, recognising that there were exceptions, the manufactured gas industry's performance through to the Second World War was reasonably satisfactory. But in the post-war period life changed. Oil was king and the major growth fuel, both the manufactured gas and coal industries saw their markets being eroded by the rapid inroads of oil for a whole variety of industrial applications, and by the increasing use of both oil and electricity within the home.

Manufactured gas was becoming uncompetitive and was regarded increasingly by the public as an old-fashioned form of fuel. In many countries, sales, and hence revenues, began to decline and the longer-term viability of various gas companies was suspect. In the late 1950s gas-making processes based on oil feedstocks appeared to offer the prospect of reversing this trend and indeed did so temporarily in the UK (see Chapter 2), but whether these would have become the long-term saviour of the gas industry, or just a short-term reprieve, is open to question. Undoubtedly the saviour of the gas industry was the arrival in the 1950s and 1960s of natural gas supplies in Western Europe, Japan, Australasia and elsewhere.

The manufacture of gas, but this time in the form of a high calorific value gas compatible with natural gas, will undoubtedly make a come-back. It will be needed to supplement declining indigenous supplies of natural gas in the years ahead, to assist in load balancing and so on, but this is another story.

The main thrust of this book is directed at the role of natural gas, how the business has developed, its special features and characteristics, and related subjects. But before leaving the subject of the manufactured gas industry, it is necessary to give recognition to one fundamental point which can all too easily be overlooked. In essence the manufactured gas industry was, as the name implies, heavily orientated towards the manufacture of gas and with the distribution of a low calorific value fuel at low pressures over relatively short distances. For gas companies the advent of natural gas represented far more than just converting consumers' appliances to accept a higher calorific value gas of different burning characteristics: it was a much more radical change for them. It necessitated a wholesale change of thinking and attitudes, in particular highly developed, long-established skills in the manufacture of gas were no longer required. Management and employees had to learn new, but not necessary any less complex, skills and techniques of how to handle, distribute, and most importantly, sell, a quite different form of gas. A whole new infrastructure had to be created virtually overnight. This was not just the creation of high-pressure, country-wide transmission systems, compressor stations and all the other necessary equipment and techniques to handle natural gas, most of all it was the people employed in the industry who had to relearn their business. The fact that natural gas has been introduced so successfully to such good effect in so many countries in recent years, reflects full credit on the rapidity with which those concerned have been able to adapt themselves to these new challenges with the minimum degree of disruption to the public at large.

2: The United Kingdom

The United Kingdom merits a chapter to itself for several reasons. First, as already described, it is generally recognised that the first commercial use of gas was in the UK in the early part of the nineteenth century and that an Englishman, among others, can claim having discovered a process to produce a combustible gas from coal over a hundred years earlier. Second, in the late 1950s the UK led the world, together with the Japanese, in pioneering the development of processes to make gas from oil feedstocks. Third, at about this time the UK also pioneered, in conjunction with American interests, the transportation of natural gas in liquefied form by ocean tanker and in October 1964 became the first country in the world to import liquefied natural gas on a long-term, commercial basis. Fourth, the British Gas Corporation is now the largest gas entity in the western world when measured in terms of the number of customers (i.e. 14.5 million) it services and supplies with gas for its own account.

The UK gas industry, can thus lay claim to a number of notable 'firsts' and to having innovated some important developments which have had a far-reaching effect on the world's gas industry as we know it today.

The Birth of the Gas Industry

As in almost all long-established gas markets, gas manufactured from coal was first used primarily as a means of illuminating streets, houses and factories. The first company to be established was the London and Westminster Gas Light and Coke Company in 1812, which had statutory powers to make 'inflammable air' and to distribute it through pipes for lighting purposes. In the following years, other companies began to make and supply gas for lighting both in London and in the provincial towns. Some of these companies operated with statutory powers, while others had no statutory powers and depended on local good will for permission to install their pipes in public streets. Many companies operated in direct competition with one another as it was not public policy at that time to grant gas companies the exclusive franchise for a designated area.

Competition between the companies was often intense, and relations with consumers were difficult as there were no meters to measure gas of uncertain pressure and indifferent and varying quality. Nevertheless, in spite of these drawbacks, gas increasingly displaced oil and candles for lighting.

By 1830, 200 gas undertakings existed. The gas meter was invented and gas began to be sold by volume instead of under a 'gas rental' basis according to the size of burner and hours of use. At the same time new techniques helped to improve gas quality and to reduce costs. Direct competition gave place to monopolies with appropriate provisions for consumer protection. Following the lead of the Manchester Corporation, municipal authorities began to go into the business, either by purchasing private gas companies, or by starting up new undertakings for their own account. By the middle of the nineteenth century gas was being used for lighting in all the larger towns and cities.

The first definitive statistical evidence became available in 1882 when the

2.1 An unusual suspended gas-fired radiator at the Royal Academy, London

Board of Trade began to collect returns from statutory undertakings, i.e. undertakings with powers from Parliament to make and supply gas in a defined area. In that year statistics reveal that there were 500 statutory undertakings with a capital investment totalling some £50 million supplying 67 000 million cubic feet. Approximately two-thirds of the gas was supplied by privately owned companies and one-third by municipally owned undertakings, a ratio which was broadly maintained until the gas industry was nationalised in 1949.

Expansion and Competition

In the 1870s the introduction of the electric dynamo, based on an invention by Hippolite Pixii in 1831, posed a real competitive threat to gas lighting. In 1882, the Electric Lighting Act led to the establishment of electricity undertakings on a statutory basis, but because of the high cost of electricity and the introduction of the incandescent gas light mantle in 1877, a vast improvement on the old open-flame burner, it was some years before electric lighting had a real impact on gas sales for lighting purposes. Meanwhile, gas began expanding into new markets such as cooking, water and space heating, and was also used for certain industrial heating purposes. The invention of the prepayment meter around this time also helped to spread the use of gas by the poorer sections of the community who could not afford to pay quarterly bills.

Between 1882 and 1912 the number of statutory gas undertakings rose from 500 to 826, largely due to non-statutory undertakings receiving statutory status. Over this period total capital investment increased to £137 million, sales to nearly 200 000 million cubic feet, and the total number of consumers rose from 1 972 000 to 6 876 000.

Before 1920 gas was sold by volume, but legislation then imposed on statutory gas undertakings the obligation to sell gas by heat content and not by volume. The unit adopted in the UK was the therm, equivalent to 100 000 British Thermal Units (Btu)—1 Btu being the amount of heat required to raise the temperature of 1 pound of water by 1 degree Fahrenheit. This meant that each gas undertaking had to declare the heat content or calorific value of its gas, in particular the number of Btu produced by the combustion of 1 cubic foot of gas measured under standard conditions of temperature and pressure.

The First World War

Up to the First World War the expansion of the industry was steady and sustained. Sales to the domestic market predominated, although sales to the industrial market were beginning to become important. Gas for indoor lighting was being displaced by electricity, notwithstanding the incandescent mantle, but even so gas was still used extensively for illumination, while competition from electricity for other domestic applications was negligible. The First World War brought many problems. The gas industry had difficulties in maintaining the statutory illuminating quality of gas because of a shortage of coal of a suitable quality, exacerbated by the extraction from gas of benzole for war purposes. In the latter case toluol could be made from benzole and toluol was, of course, required for the manufacture of tri-nitrotoluene, or TNT as it is more commonly known. Benzole was also used as a fuel for motor vehicles.

Other developments during the early 1900s were the increasing utilisation of the by-products from gas making, notably coke and tar. Coke gradually became an acceptable fuel in its own right, while tar was used for both road surfacing and as a feedstock for the manufacture of many high-value substances including aniline dyes.

Between the Wars

Between the First and Second World Wars the gas industry continued to expand, but at a decreasing rate as competition from electricity made itself felt. Gas sales between 1920 and 1929 rose from 1177 to 1465 million therms, but in the next few years were below the 1929 level and did not rise above it until 1935. During the period 1920–38 the number of gas undertakings declined from 798 to 703, due to the process of amalgamation, while the total number of consumers increased from 7 448 000 to 11 215 000. However, the average consumption per domestic consumer showed a steady decline because, or so it would seem from statistical data, the majority of new domestic customers that were acquired were prepayment consumers implying that they comprised, in the main, consumers with a limited purchasing power. Other factors which contributed to this reduction in average consumption were the improved efficiency of appliances, the continued loss of the lighting load to electricity, and competition from electricity for cooking and heating.

The gas industry recognised the seriousness of this downward trend in consumption and endeavoured to counter this with improved service, salesmanship and publicity. But these efforts were not unified over the country and their effect was rather limited. At the same time many undertakings sought to sell more gas to industry to offset their domestic losses.

The Second World War

The heavy air raids of the Second World War caused substantial destruction of manufacturing, storage and distribution plant, and seriously affected output and sales of gas. There were, however, some compensations. Demand

2.2 The retort house in an early manufactured gas plant at Pinner, Middlesex, England, using coal as a feedstock (*British Gas Corporation*)

for gas rose in areas not subjected to heavy bombing, particularly for the new war industries that were established in such areas. Domestic consumers also found gas very valuable as solid fuel supplies became short. In fact between 1939 and 1944 sales by statutory undertakings rose from 1487 to 1737 million therms. Demand for the by-products of the gas industry such as benzole, tar, and hydrogen (for barrage balloons) all rose substantially in these war-time circumstances.

Committee of Enquiry

On 12 June 1944, the Minister of Fuel and Power appointed a five-man committee of enquiry into the gas industry headed by Mr Geoffrey Heyworth (later Lord Heyworth) with the following terms of reference: 'To review the structure and organisation of the gas industry, to advise what changes have now become necessary in order to develop and cheapen gas supplies to all types of consumers, and to make recommendations.'

The committee completed its report in November 1945 and this was presented to Parliament by Emmanuel Shinwell, Minister of Fuel and Power, the following month.

In essence the committee of enquiry recommended, inter alia, the compulsory purchase (by government) of all existing gas undertakings with fair compensation, and the creation of nine regional boards in England and Wales and one for Scotland. This recommendation to take the gas industry into public ownership accorded with the policy of the Labour government which had taken office for the first time with a parliamentary majority in July 1945.

At the time of the Heyworth Enquiry, the industry comprised 1047 undertakings, which ranged in size from undertakings supplying single villages to the (London) Gas Light and Coke Company, which alone accounted for 12 per cent of total gas sales. In addition, there were 19 companies, such as railways and collieries, supplying small quantities of gas to the public as an activity subsidiary to their main business. Total sales (excluding Northern Ireland) at that time were 1756 million therms, while the number of gas consumers amounted to about 10.5 million.

Whether one supports or not the concept of public ownership, and whether one agrees or disagrees that a state-owned monopoly is the most effective way of running the gas industry is, of course, a matter of personal opinion. However, few people would disagree that in the mid-1940s the UK gas industry with its multiplicity of undertakings—334 of which had a capital of less than £20 000—its varying standards of gas quality and service, etc., would benefit from some measure of rationalisation and the establishment of larger, more efficient entities.

Nationalisation

Unlike the colliery and electricity companies—industries which were by this time already nationalised—the gas companies did not embark on a campaign of opposition to nationalisation. Perhaps they thought it would be futile after other more powerful industries had been taken over, and after the revelations of the Heyworth Report they could not really put up much of a case. It should also be noted that the coal industry, on which the gas industry depended for its feedstock, had been nationalised, likewise its major competitor, the electricity industry. This gave these two important industries state backing and access to cheap government-guaranteed capital.

In any event the gas industry put up no worthwhile opposition to a state take-over and little interest was shown by the general public. The only opposition was in Parliament, where the Conservative Opposition put up a fierce fight. Indeed the final session of the Gas Bill lasted without a break for a then record period of 51 hours.

In 1948, Parliament decided to nationalise the gas industry and this took place in 1949. At that time there were 1050 gas undertakings in existence, supplying a total of 2119 million therms to 11.3 million consumers. The Gas Bill provided for the establishment of 12 autonomous area boards, not 10 as proposed by the Heyworth Committee, which were responsible for arranging their own supplies and finance. The latter, and certain other matters such as industrial relations, had to be arranged through a central organisation. For this the Bill provided for a Gas Council consisting of the 12 area board chairmen, together with an independent chairman and a deputy chairman. The Gas Council had a small organisation, and except on matters of general policy, the area boards proceeded independently of each other.

Modernisation

The gas industry then set about modernising itself, a not inconsiderable task after the destruction caused by the war and the lack of preventive maintenance during that time. In the 11 years from nationalisation in 1949 to March 1960, 622 old and inefficient gasworks were closed, some of the larger works were extended and linked together, and 21 200 miles of new mains were laid.

In spite of the considerable sums spent on rebuilding plant and mains, domestic consumption declined from 1366 million therms in 1953 to 1268 million therms in 1960. There was, however, an increase in industrial consumption, which rose from 639 to 819 million therms over the same period. One of the problems was the increasing cost of and difficulty in obtaining suitable supplies of coking coal, which came largely from older coal fields in Durham and from Yorkshire, and for which the steel industry was a ready customer. In the 10 years from 1950 to 1960 the average price to the domestic consumer rose from 14.8 to 24.91 old pence per therm, an increase of 65 per cent. Over the same period the comparable price of electricity rose from 1.346 to 1.687 old pence per unit, an increase of only 25 per cent.

New Processes

The decline in domestic consumption, the increasing cost of coal, and the difficulties the gas industry were experiencing in selling coke, which was potentially an important source of revenue, obliged the industry to investigate alternative and more efficient processes to carbonise coal. One alternative was the total gasification of low-grade coals by the Lurgi process. In the mid-1950s two Lurgi plants were built in Scotland and the Midlands, but these did not prove to be as satisfactory or as cheap as was hoped. By the late-1950s attention, therefore, turned to using oil as a feedstock. Initially this took the form of siting gasworks near oil refineries which received so-called refinery tail gas, an unrefined and varying cocktail of combustible gases, and reformed it into town gas. But by the early 1960s various processes were developed by Shell, ICI and the Gas Council for transforming oil, mainly naphtha-type fractions, into gas—it was from these developments that the generic name of gasoil was derived. These oil-based processes were much cheaper than coal carbonisation, some £10 per therm plant cost for the former, compared with about £60 per therm for the latter. Moreover, oil-based plants were smaller than coal plants for a given output, were more acceptable environmentally, and produced non-toxic gas.

In spite of the Labour government's desire to maintain outlets for coal, the case for switching to oil as a feedstock for gas making was abundantly clear and was recognised by the government in a White Paper on fuel policy issued in 1965 which stated: 'The government is satisfied that the trend of the gas industry towards petroleum should be accepted. Measures to influence coal consumption by the gas industry (other than strictly limited adjustments from year to year) would involve a heavy economic penalty, and would not be in the interests of the economy as a whole.'

Nevertheless these oil-based processes were not without their problems in that they were technically complex, required special steels for their construction, and had to be operated at high pressures and temperatures. In spite of these difficulties these new processes put the gas industry back into a competitive position and a concerted sales and promotion campaign using the theme of 'High-Speed Gas', combined with the availability of more sophisticated gas appliances, led to a speedy growth in gas-fired central heating. In general it can be said that the switch from coal to oil was highly successful and gas became a growth business once again.

Liquefied Natural Gas

Paralleling these efforts in time, the Gas Council embarked on a serious investigation into the importation of liquefied natural gas, involving a series of trial shipments from Louisiana to Canvey Island in the Thames Estuary between 1959 and 1960. These trials were successful and in November 1961 the government authorised the Gas Council to enter into a 15-year contract to purchase 350 million therms per year (about 100 million cubic feet per day of

2.3 One of the modern traditional coal-gas manufacturing plants of the North Thames Gas Board at Beckton, east London, England, prior to the conversion of the UK to natural gas. Batteries of coke ovens are seen on the left (*British Gas Corporation*)

high calorific value natural gas) from Algeria. The first delivery against this contract was made in October 1964 and it marked the start up of the world's first commercial international LNG project (further background information on this project is given in Chapter 7).

It is interesting to record that this contract was concluded by the Gas Council on behalf of 8 of the 12 area boards, who were supplied via a new high-pressure methane grid which ran from Canvey Island to as far north as Manchester and Leeds. The area boards in question could utilise the gas as they wished, but for the first time the Gas Council was responsible for this co-operative scheme, whereas hitherto each area board had autonomous responsibility for producing and selling its own supplies of gas. This was a significant development and presaged future changes in the structure of the industry.

The situation had now been reached where coal-based gas was being increasingly displaced by oil-based gas and supplemented by imports of high calorific value natural gas in liquefied form—the latter being reformed down to town-gas quality of approximately 500 Btu per cubic foot before it was distributed to end-consumers.

North Sea Gas

The next significant milestone was the Continental Shelf Act of 1964 which vested in the Crown all the rights in the UK sector of the North Sea and enabled the Crown, through the Ministry of Power, to grant licences to explore for hydrocarbons in this area.

The Continental Shelf Act was, of course, to enable effect to be given to certain provisions of the international Convention on the High Seas initiated in Geneva in April 1958. This convention was to come into force after ratification by 22 nations; Britain was the 22nd nation to do so on 10 June 1964. In brief, the Geneva Convention gave coastal nations rights over the natural resources below the sea bed up to a water depth of 200 metres—or deeper if mineral exploitation was possible. Apart from a deep trough off the Norwegian coast, most of the North Sea is less than 200 metres deep.

Most of the states around the North Sea accepted that their boundaries with their neighbours should be settled on the basis of median lines equidistant from their respective coasts. However, there was a dispute between West Germany, Denmark and the Netherlands and a different boundary was agreed between these three countries in 1969.

In September 1964 the government granted the first exploration licences in the UK sector, covering 348 blocks to 22 companies or consortia, who embarked on an exploration programme concentrated on the southern sector of the North Sea. The first success was achieved by British Petroleum when it discovered the West Sole gas field in October 1965. This was followed by the discovery by Shell/Esso of the Leman field in April 1966; Indefatigable (Amoco) and Hewitt (Arpet) also in 1966; Dottie (Phillips) in 1967; Deborah (Phillips), Rough (Amoco) and Viking (Conoco) in 1968; and South-East Indefatigable (Shell/Esso) in 1969.

These discoveries—all non-associated gas fields—were against the background of the Continental Shelf Act provisions in which:

> The holder of the licence shall not without the consent of the Ministry of Power use the gas in Great Britain and no person shall without that consent supply the gas to any other person at premises in Great Britain.

and

> The Minister of Power shall not give his consent . . . to the supply of gas at any premises unless satisfied—that the supply is for industrial (i.e. non-fuel purposes in this context) purposes and that the Area Board in whose area the premises are situated has been given an opportunity of purchasing the gas at a reasonable price.

These and other provisions of the Act in effect meant that to all intents and purposes any gas discovered in the UK sector had to be offered to the Gas Council/area boards; what would be considered as a 'reasonable price' was not defined. It was also quite evident that no licence holder could attempt to sell any gas discovery he might make to other potential buyers, for example in continental Europe, until such time, if ever, as the UK's requirement for gas was more than satisfied. Exploration, therefore, went ahead against a background of uncertainty as to what rewards (price) were likely to be forthcoming for the exploration risk taken, but in the certain knowledge that the explorers would have to sell any gas they found to the state-owned entity, namely the Gas Council acting on behalf of the area boards.

The first sale by BP to the Gas Council for West Sole gas was concluded on relatively attractive terms by the standards of that time, but the price attached to this sale was for an initial three-year period only. In retrospect it appeared to serve its purpose of encouraging other explorers to maintain their efforts. However, all subsequent deals, up to the conclusion of the Frigg gas deal in 1973, were substantially below the initial West Sole terms. As in all situations of this nature, it is impossible to say beyond dispute whether the exploration effort would have been substantially greater if the long-term contracts concluded in the late 1960s had been made on terms more attractive to the producers. Such evidence as there is indicates that the exploration effort could have been greater, in that in the fourth round of allocations in August 1971 of the 79 blocks offered in the southern sector only 21 were taken up; attention had by then turned to the potentially more interesting oil prospects in the northern sector.

Contractual Terms for North Sea Gas

The reader will note that the author has not quoted any contract (delivered at the beach) prices for the original three-year West Sole deal, or for the subsequent long-term deals for other southern and northern North Sea gas contracts to illustrate the foregoing observations. Not unreasonably—indeed it is customary practice in most countries outside North America—contractual details, in particular pricing arrangements, are confidential matters as between buyers and sellers.

However, British Gas has published some information on the broad scope and concepts adopted in formulating its supply contracts with the producers. In summary, the main contracts originally concluded for southern North Sea gas were for 25 years' duration and of a 'depletion' type nature. For the larger fields the depletion rate was set so that in a typical full production year 1/20th of the total reserves would be produced, i.e. an average daily production rate of 1:7300, but for some of the smaller fields a faster depletion rate was fixed for economic and other reasons.

The contracts provided for variations in offtake as between summer and winter, not only to take account of changes in the seasonal market demand pattern, but also to allow for maintenance work on wells, platforms and other facilities to be carried out. Typically the load factor for these southern sector contracts was 60 per cent. Although on a daily basis the buyer had the right to take gas within a fairly wide range, with agreed minimum and maximum rates, over the year as a whole the Annual Contract Quantity (ACQ) must be taken, or if not taken must be paid for—the so-called 'take-or-pay provision' which is a quite common feature of many gas contracts throughout the world. On the other hand gas not taken, but paid for, could normally be taken (in the case of these southern North Sea contracts) free of charge at some later date once the ACQ for that year had been taken. Also gas taken in any one contractual year in excess of the ACQ was paid for at a lower price. In the UK this lower-priced excess gas has become to be known as 'valley gas'.

Touching on price, the Gas Council/British Gas have stated that the prices originally negotiated were neither 'cost-plus' nor 'market-related' prices, and that price-review mechanisms for these original contracts concluded at a time of relatively low rates of inflation were fairly rare. There were, of course, many other contractual provisions concerning such matters as gas quality, determination of reserves, nomination, payment, arbitration procedures and so on, all of which impinged to a lesser or greater extent on the overall worth of the total contractual packages that were concluded between the sellers and the buyer.

The situation so far described relates essentially to those contracts settled in the late 1960s to early 1970s for non-associated gas from the southern basin. Because of the dramatic changes that have taken place since then as regards the overall economic climate, the more recent concluding of contracts for associated gas, where gas availability is conditioned by oil production rates, and the influence of competitive bidding with potential continental European buyers for Frigg gas, modifications are being made to the original sourthern-basin contracts. These modifications are taking the form of greater flexibility in the quantities of gas that British Gas are obliged to take, extended contractual periods, and similar easements for the buyer which will help to facilitate the accommodation in British Gas's overall supply/distribution system of northern North Sea gas—including the increasing contribution expected from associated gas with its inherent production characteristics—in the years ahead. In return for these modifications and for the new investment that the producers will have to make, compensatory price improvements were negotiated in 1978. However the resultant prices are still low by international standards.

Conversion

The extent of the discoveries made in the southern sector was such that it opened the door for the UK to abandon manufactured gas and to convert the whole country, or more precisely England, Wales and Scotland, but not Northern Ireland as yet, to natural gas. The more recent discoveries in the northern sector, mainly in this case of associated gas, have reinforced the rightness of this decision. Let us, therefore, turn our attention to the story of conversion.

With the security of substantial supplies of natural gas under long-term contracts, the Gas Council decided that the best way to maximise the benefits of natural gas was to convert existing appliances to use natural gas, rather than convert, i.e. reform and down-grade, the gas to suit the appliances. In retrospect, and in the light of the experience gained in North America, the Netherlands and elsewhere, this was the obvious decision to take. Nevertheless, it was a decision not taken lightly and the alternatives of down-grading natural gas to existing town gas, or to some half-way standard between town gas and natural gas quality, were seriously considered. Before the final decision was taken a pilot conversion scheme was carried out in the Canvey Island area in 1966. This was considered to be sufficiently successful to decide to embark upon country-wide conversion in 1967.

In taking this decision the Gas Council had to face the fact that to the estimated conversion cost of some £400 million would have to be added the cost of writing-off prematurely over £400 million that was invested in gas-making plants. Many of the latter were quite new, or indeed still under construction, particularly those that were designed to work on oil feedstocks.

An important factor which influenced the decision to convert was that the expansion in gas demand had already begun to place a strain on the industry's storage and distribution facilities. Natural gas with double the calorific value of manufactured gas offered the prospect of putting a great deal more energy through the existing distribution system. The alternative of building new distribution mains, additional reforming plants and storage facilities could well have cost some £2000 million as compared with the estimated £800 million for conversion and plant write-off.

Apart from the costs of conversion and the writing-off of obsolete plant, the introduction of natural gas necessitated investing over £2000 million in coastal reception terminals, high-pressure pipelines, compressor stations, control systems, pressure reduction stations, off-take points, and in

2.4 Conversion activities in a Midlands town in England. Manufactured gas is burnt off at this purging point as it is pushed out by the in-coming natural gas. A change in the flame indicates when natural gas has filled the supply system (*British Gas Corporation*)

2.5 A large diameter, high-pressure natural gas pipeline being laid in Cheshire, England. Once the work is complete the pipe will be buried and the land and river banks restored to their original condition (*British Gas Corporation*)

reinforcing and adapting local distribution systems. However, much of this expenditure would be incurred whether the conversion or the reforming route was adopted and thus this expenditure did not materially affect the decision either way.

The task before the industry was huge by any standards. There were some 13½ million premises to visit, not once but two or three times at least; 35 million appliances, with perhaps 200 million burners, and 8000 different domestic appliance models to convert. Although conversion is always an inconvenience to the consumer, it gave the area boards a unique opportunity

2.6 Indefatigable gas field, southern North Sea, UK sector. The 1200 tonnes deck section being lifted on to the jacket of Shell/Esso's 'L' production platform by the derrick barge *Odin*, 15 April 1978 (*Shell*)

to find and rectify many appliances which were in poor condition, incorrectly installed, or inadequately ventilated. Appliances were overhauled and when they could not be converted safely and efficiently, were either replaced free of charge or new appliances were made available on advantageous terms. Conversion, therefore, facilitated the carrying out of a major upgrading in safety standards throughout the country.

Unlike most other countries which had already converted, the UK decided to supply sets of replacement parts that could be fitted to an appliance in situ. Elsewhere the usual technique had been to remove the appliance from the home and then to modify it in mobile workshops in the locality. This new approach certainly helped to speed up conversion. This was finally completed in September 1977, a somewhat shorter period, in relation to the number of customers and appliances involved, than had been achieved in most other countries.

TABLE 2.1 PROGRESS OF CUSTOMER CONVERSION TO NATURAL GAS

Year (April–March)	*Total customers*	*Annual number converted to NG (000s)*	*Cumulative number converted to NG*
1967/68	13 210	51	51
1968/69	13 265	418	469
1969/70	13 347	1 093	1 562
1970/71	13 372	2 029	3 591
1971/72	13 390	2 407	5 998
1972/73	13 506	2 100	8 098
1973/74	13 559	2 108	10 206
1974/75	13 682	1 674	11 880
1975/76	13 925	1 131	13 011
1976/77	14 200	329	13 340
1977/78	14 516	98	13 438

Note: Total customers include those customers acquired during the conversion programme, who used natural gas right from the outset. As stated above, all customers at one time on manufactured gas had been converted by September 1977.

The final cost of conversion was reported to be £577 million which, British Gas claim, corrected for inflation was less than the estimate of £400 million made in 1966. To this has, of course, to be added the £450 million write-off of obsolete plant, bringing the total bill to £1027 million. British Gas further stated in their Annual Report and Accounts for the financial year April 1977/March 1978 that: 'The Corporation finished writing off the whole of this sum (i.e. £1,027 million) during the year under review. The cost of the entire conversion programme was met without external subsidy of any kind.'

There are some, especially the private industry producers of North Sea gas, who would probably dispute the latter part of this statement. They would claim, with some justification, that the low prices they received at the beach for natural gas supplies constituted a substantial indirect subsidy. Certainly, since the latter part of the 1960s, the prices received by the producers from British Gas for gas from the southern sector of the North Sea have been among the lowest (for comparable production conditions) for any gas market of consequence in the western world. Evidence for this can be derived from British Gas's Annual Report and Accounts for the financial year April 1976/March 1977 which indicates that British Gas paid an average of approximately 1.9 pence per therm for natural gas at the beach during this period. This compares with British Gas's average retail selling prices as given in *Table 2.2*:

TABLE 2.2 AVERAGE PRODUCER AND RETAIL PRICES

Average prices: April 1966/March 1977	*Pence per therm*	*US¢ per MMBtu*
Ex-beach price to producers	1.9	32
Retail price to —domestic consumers	16.65	284
—industrial consumers	7.24	124
—commercial consumers	14.00	239
—average all consumers	12.20	208

It is also interesting to note that over the same period the average selling prices in the UK of kerosene for domestic consumption were about the equivalent of ¢350 per million Btu (MMBtu), and of heavy fuel oil and gasoil to industrial users about ¢250 and ¢325 per MMBtu respectively. The margin between gas and electricity prices was even greater.

Part of the reason for the undoubted success that natural gas has achieved in recent years, in particular its rate of growth at a time when the overall economic activity of the country has been depressed, has been the result of the fact that gas has been consistently cheaper for the end-user than competitive forms of energy. In fairness it must be recorded that this pricing policy for gas has been largely controlled and directed by successive governments, and one can only speculate what British Gas might have done had they had a free hand in such matters.

However, to end the conversion story on a less provocative note, few would dispute that it was a highly successful and professional operation once the inevitable teething troubles were overcome, and that the number of incidents that occurred was remarkably small in relation to the magnitude of the task.

British Gas Corporation

The reader will have noticed several references to British Gas in the foregoing account of conversion. The Gas Act of 1972 provided for the setting-up of British Gas Corporation, which came into being on 1 January 1973. It replaced the Gas Council and 12 area boards' structure established on 1 May 1949. British Gas was given the responsibility 'to develop and maintain an efficient, co-ordinated and economical system of gas supply for Great Britain, and to satisfy, so far as it is economical to do so, all reasonable demands for gas in Great Britain'.

The Act provided for the appointment by the Secretary of State for Energy of not less than 10 nor more than 20 corporation members in addition to the chairman. The corporation was empowered to determine the industry's operational structure and to exercise its responsibilities through twelve regions with the same geographical boundaries as the original area boards. Each region has its own chairman and deputy chairman appointed by the corporation.

This change was considered necessary as a direct result of the discovery of natural gas in the North Sea necessitating the conduct of the industry's affairs, in particular the acquisition and distribution of natural gas, on a nation-wide basis; it was no longer practicable or appropriate, in the government's opinion, for area boards to have autonomy on such matters. For a monopolistic, state-owned industry this change was probably inevitable in that it eliminated any possibility, longer term, of area boards competing with each other for supplies, if a situation should develop where demand began to exceed uncommitted supply availability. In practical effect such a possibility was unlikely to arise, as evidenced by the fact that the Gas Council negotiated the purchase of Algerian LNG on behalf of eight area boards and was the focal point for negotiating North Sea gas supplies before British Gas came into being. Nevertheless, the creation of British Gas tidied up a situation which hitherto was not strictly speaking enshrined in the powers originally granted to the Gas Council.

Progress Achieved between 1967 and 1977

The gas industry has changed radically in all manner of ways since the last war. For illustrative purposes comparison is made between the financial years 1967/68 and 1976/77 based on statistics published by British Gas. This 10-year period has been selected as it embraces the fundamental change from manufactured gas to natural gas. Latest available data for the year April 1977/March 1978 are given in *Table 2.5*

First let us look at how the pattern of gas manufacture and supply availability has changed (*Table 2.3*).

TABLE 2.3 RAW MATERIALS USED FOR GAS MANUFACTURE/SUPPLY

Materials	*Units*	*1967/68*	*1976/77*
Coal	Thousand tons	13 606	8
Oil	Thousand tons	4 917	92
LPG	Thousand tons	924	26
Natural gas	Million therms	662	14 694

And second, how the outputs of various main types of gas-making plant have fallen away over this period (*Table 2.4*).

TABLE 2.4 TYPES AND CAPACITIES OF GAS-MAKING PLANTS (MILLION CUBIC FEET DAILY)

Type of plant	*1967/68*	*1976/77*
Total carbonising	646.6	0.6
Lurgi process	96.0	–
Oil gasification—cyclic process	1 003.0	131.5
—continuous process	3 199.3	
Water gas and others	806.8	232.0
TOTAL	5 751.7	364.1

The number of gas works declined over this period from 192 to 35. The effect of this on the number of people employed by British Gas (or by its predecessor the Gas Council and the 12 area boards) was quite dramatic. In 1967/68, manual workers employed on gas-manufacturing activities amounted to 24 358 out of a total establishment of 122 784; 10 years later the number had dropped to 1474 (out of 99 926).

Total customers (domestic, commercial and industrial) rose from about 13.2 to 14.2 million, the increase of approximately 1 million being entirely attributable to the domestic sector. Average consumption per domestic customer had meantime more than doubled from 211 to 453 therms per year. Annual sales of new domestic appliances remained remarkably constant over this 10-year period with one significant exception. For cookers, sales averaged about 600 000 per year, water heaters nearly 200 000, and space heaters over 900 000. The exception has been sales of central-heating units which increased from 283 000 in 1967/68 to 508 000 in 1976/77. Within this total, warm-air units have lost popularity and sales have fallen progressively each year from 106 000 to 40 000, while sales of gas-fired boiler units have increased from

177 000 to 468 000. Average unit proceeds from gas sold to domestic customers have risen from 10.29 to 16.65 pence per therm, the largest increases being in the last two years.

Although the number of industrial consumers in fact declined from 78 000 to 69 000, annual sales to this market sector expanded from 915 to 6107 million therms, the largest single increase being in respect of sales to the chemical industry—from 36 to over 2000 million therms per annum in 10 years. The number of commercial consumers also declined, but total sales to this class of market increased steadily over the years, i.e. from 633 to 1547 million therms. Published statistics on average unit proceeds from industrial and commercial consumers are less meaningful than is the case with the domestic market. This is because the tariffs for the former two classes of market are obviously influenced by substantial differences in the volumes supplied to individual consumers, load-factor variations, whether the sales in question were on a firm or interruptible basis, and similar contractual considerations.

Financial Aspects

Analysis of the financial results of British Gas (and its predecessors) is far more complicated. In round terms, gross turnover in 1967/68 was £590 million, of which the cost of prime materials (i.e. coal, coke, oil and purchased gas) constituted £195 million or nearly 33 per cent. In 1976/77 turnover was £1958 million, of which the cost of prime materials (almost entirely purchased natural gas) amounted to £285 million or nearly 15 per cent. For the two years in question—there were variations up and down, of course, in the intervening eight years—salaries and wages, other operating costs and interest payments showed little variation when measured in terms of the percentage of gross turnover; these being approximately 22, 30 to 27, and less than 10 per cent respectively. The main item that has changed substantially, other than the aforementioned cost of prime materials, has been depreciation and write-offs. The latter have increased from less than 8 per cent in 1967/68 to 26 per cent in 1976/77. This reflects, inter alia, British Gas's current policy of accelerating the rate of write-off of the costs of conversion and redundant gas-manufacturing plant.

Against this background, and during a period of fundamental change from manufactured gas to natural gas, trends in profitability become distorted and cannot be analysed in a meaningful way. All one can do is to record the facts. In 1967/68 the consolidated result of the Gas Council and the 12 area boards was a loss of £12.9 million. In each of the following five years, profits were made ranging from £1.1 million to £17.5 million. In the years 1973/74 and 1974/75 losses were made of £41.3 million and £30.8 million respectively. But for the last two years of the period British Gas was again in the black with profits of £25.1 million and £31.5 million respectively. For 1976/77 the net return on capital employed before interest amounted to 9.3 per cent, and after interest to 1.3 per cent. With conversion now completed and with the costs of this and of redundant plant fully amortised (in March 1978), it will be interesting to see how the profitability of British Gas develops in the years ahead. To the extent that the return on net assets exceeds 10 per cent in any one financial year, then the Gas Act of 1972 provides for any such excess to be paid over to the Treasury, subject to certain caveats.

Finally, before leaving matters of a financial nature, British Gas's self-financing ratio has moved from 17 per cent in 1967/68 to 100 per cent in 1976/77. In this regard the Gas Act limits the corporation's borrowing powers to £2700 million which by March 1977 had been exercised to the extent of £2100 million. Of this latter amount some £1440 million constituted government advances, mostly at 8.1 per cent average interest rate, and £414 million of foreign loans, two-thirds of which are at an average interest rate of 10.6 per cent.

The Contribution from the Private Sector and Other Bodies

Inevitably, this account of the development of the gas industry since the Second World War has concentrated on the activities of the Gas Council and British Gas. Perhaps insufficient attention and credit have been given to the essential complementary, pre-requisite efforts by private companies, including such state-owned entities as British Gas, in exploring for, producing, and bringing natural gas from the North Sea to the UK beach. This has been an outstanding achievement by any standards in what is, all said and done, one of the world's most hostile, technically complex, and high-cost environments that the industry has had to operate in to date. In the more northerly areas during winter the wind force is above 17 knots for 60 per cent of the time. In an average storm a maximum wave of 20 metres can be expected and wind gusts of over 100 knots. Air temperatures can be below freezing for long periods. Although conditions are not quite so severe in the mainly gas-bearing southern sectors of the North Sea, they nevertheless have been a considerable challenge in their own right. However, as the story of the discovery and development of the North Sea's oil and gas reserves has been recounted many times already, it does not require further repetition here.

In concluding this brief account of the progress made so far, recognition should also be made of the many other organisations who have played their part in transforming the UK's gas industry. Construction companies and contractors who have built and installed production facilities, pipelines, reception terminals, compression stations and all manners of necessary equipment; appliance and conversion-kit manufacturers; the role played by banks and other institutions who have provided the necessary financial

2.7 Semi-submersible pipe-laying barge *Semac I* laying a 36 inch, 280 mile gas line from the Brent Field to St Fergus, Scotland, UK sector of the North Sea, March 1977 (*Shell*)

resources, particularly for the smaller companies with limited resources of their own—the list is virtually endless. Nor should one forget the part played by governments, both Conservative and Labour, who at least in the early days of North Sea development were expeditious in the allocation of the necessary exploration licences in order to get the show on the road as quickly as possible. All deserve their appropriate share of praise for the contributions they have made, be they of a primary or of a supportive nature.

Future Supply Prospects

Supply availability from existing contracts for non-associated gas in the southern sector of the North Sea reached a peak average rate of some 4000 million cubic feet per day (MMcf/d) in 1977. For the remainder of the 1970s and through to the early 1980s, production is expected to fall to an average plateau rate of around 3500 MMcf/d, before entering a slow but progressive decline. By 1990, average annual supplies from these southern fields (i.e. West Sole, Leman, Indefatigable, Hewett, Dottie, Deborah, Viking and Rough) could be down to around 1500 MMcf/d or perhaps somewhat lower.

British Gas also has under firm contract both UK and Norwegian sector non-associated Frigg gas, the latter secured in the face of competition from continental European buyers. Frigg came on stream in the autumn of 1977 and supplies are scheduled to build up to a rate of about 1500 MMcf/d by 1980 and to stay at this level until about 1988, falling to around 1000 MMcf/d by 1990. To this can be added at least a further 500 MMcf/d of associated gas from the northern Brent oil field with initial supplies commencing in 1980. And it is also possible that British Gas may take up its option with Sonatrach for the continued import of 100 MMcf/d of Algerian LNG when the present contract expires in 1979.

On the foregoing basis, existing contracts represent a total known supply availability rising from just over 4000 MMcf/d in 1978/79 to around 5500 MMcf/d in the early 1980s, declining thereafter to perhaps 3000 MMcf/d by 1990. However, known non-associated gas reserves in the southern North Sea, which have yet to be committed to market, plus British Gas's Morecambe field in the Irish Sea, and the linking in of certain associated gas fields to the existing Brent and Frigg pipeline systems, should largely offset this decline in existing contracted-for supplies. Collectively these prospective supplies, together with existing contracts, should give British Gas a total average supply availability of between 5500 and 6000 MMcf/d throughout the 1980s.

Further supply prospects may materialise. In essence these comprise a variety of small associated and non-associated gas fields for which no firm development plans yet exist. Few of these fields have sufficiently large enough reserves of gas to justify individual pipelines to shore, but economic development may be possible if some multi-field gas-gathering systems can be established. Several studies are being made of this possibility; however it has not yet been decided which, if any, of several different schemes might be technically feasible and economic in their own right. In any event it is unlikely that the development of these small fields would have a significant impact on supply availability much before the second half of the 1980s.

A final possibility, apart from new discoveries, the importance of which will be conditioned by their size and geographic location, is uncommitted Norwegian gas. Several associated and non-associated gas fields lying close to the median line boundary between the UK and Norway await development. The gas reserves in question are far greater than Norway could possibly utilise itself over a long period and this gas could be sold either to continental Europe, or to the UK, or—perhaps more likely—partly to each market.

2.8 Shell/Esso's Brent 'A' oil and gas production platform in the northern UK sector of the North Sea. This photograph was taken shortly after sunrise in gale force conditions in November 1977, with wave heights of 20 to 30 feet (*Shell*)

Studies on the development prospects for such gas are also in hand, but no decisions have yet been taken. However, it is not unreasonable to assume that some Norwegian gas, over and above the Norwegian part of Frigg, will find its way to the UK.

The totality of these firm and prospective supplies, plus the possibility of some new discoveries and/or additional Norwegian gas, gives the UK gas industry, in effect British Gas, a relatively substantial supply base which at the worst may level off as from the early 1980s at about 5500 MMcf/d, but which may grow to 7000 to 8000 MMcf/d by the 1990s. What quantities will actually be realised will be conditioned by many factors, not least of all the incentives that the producers may be offered to prove up and develop new reserves as expeditiously as technical, political and economic conditions permit. Certainly, as far as uncommitted Norwegian gas is concerned, competition between the UK and continental Europe is likely to be fierce.

Demand Considerations

Turning now to the demand side, British Gas now has sufficient supply availability under contract, or in prospect, to meet the expected growth in demand of the premium market through at least to 1990 and probably longer ahead than this. In this context the premium market is taken to mean domestic and commercial consumers, plus relatively small volumes of gas for high-grade industrial applications. The growth of the premium market, which to all intents and purposes is synonymous with the so-called Public Distribution (PD) market, will not, of course, be a straight upward trending line as in practice there will be yearly peaks and troughs reflecting the effect of seasonal changes on demand. Neither will progression in demand from year to year be constant and regular. Annual growth rates will be affected by variations in rates of economic activity; in home building and modernisation programmes; in price relationships as between gas and competing energy forms; and by governmental policies which may directly or indirectly favour one particular fuel. Nevertheless, as living standards improve and the demand for space-heating grows, there can be little doubt that the PD market will continue to expand during the period considered in this assessment.

As supply availability will be greater than the PD market can absorb, this will enable British Gas to compete with oil and coal in the under-boiler and chemical feedstock markets. British Gas may well adopt a policy of supplying some of these customers on a firm year-round basis against medium-term contracts (say between 3 to 8 years), and others on a more flexible but lower-value interruptible basis. In this way, together with adequate storage arrangements, they should be able to achieve a fairly high overall load factor, while at the same time reserving the option to cut back on industrial sales if supply availability should prove disappointing and/or if the PD market should develop at a faster rate than expected.

It is not possible to assess accurately the ultimate size of the natural gas resources that remained to be discovered in UK territorial waters. The North Sea is still a fairly new exploration area and there are other areas around the UK such as the Western Approaches, the English Channel and the Irish Sea, where exploration has only recently started or has yet to begin. All one can say positively is that such resources, whatever they may be, are finite and thus, at some as yet indeterminate date, indigenous supplies of natural gas will inevitably decline. As already discussed, there would seem to be no problem in meeting the potential demand for gas for some years to come, but as an insurance for the future British Gas are researching and developing new processes for the manufacture of so-called synthetic or substitute natural gases (SNG) from coal and oil feedstocks, i.e. gas that would be compatible with natural gas. Ultimately SNG might well become the main source of

TABLE 2.5 BRITISH GAS: STATISTICAL DATA FOR APRIL 1977/MARCH 1978 (BASED ON BRITISH GAS CORPORATION'S ANNUAL REPORT AND ACCOUNTS)

	Domestic	*Commercial*	*Industrial*	*Total*
No. of customers (000s)	13 963	483	70	14 516
Sales (million therms)	6 964	1 748	6 460	15 172
Sales revenue (£ million)	1 287	287	625	2 199
Unit revenue (pence/therm)	18.5	16.4	9.7	14.5

	million therms	*per cent*
Commercial sales		
—Schools, hospitals, medical services	556	32
—Hotels, restaurants, shops	454	26
—National and local government	221	13
—Offices, banks, transport services	215	12
—Others	302	17
TOTAL	1 748	100

	million therms	*per cent*
Industrial sales		
—Pottery, glass, brick and cement works	589	9
—Chemicals and allied trades	2 207	34
—Ferrous and non-ferrous metals	757	12
—Engineering and metal goods	1 153	18
—Others, incl. textiles, food, paper, etc.	1 754	27
TOTAL	6 460	100

Sales of appliances	*000s*
—Cookers	572
—Water heaters	175
—Space heaters	917
—Central-heating units—warm air	32
—boilers	520

Mains and services—136 000 miles
Total employees—99 601

TABLE 2.5 *(contd.)*
Financial data

Assets, turnover, profits, etc.	*£ million*
—Fixed assets (net book value)	1 591.8
—Gross turnover	2 568.1
—Average capital employed	2 210.1
—Net payments to government (repayment of loans etc.)	638.3
—Cost of natural gas supplies	460.8*
—Profit before taxation	180.3
—Profit after taxation	103.9
Costs as a percentage of turnover	*per cent*
—Cost of natural gas supplies	18.2
—Salaries, wages, other operating costs	47.0
—Depreciation, write offs, interest	27.8
—Profit before taxation	7.0
	100.0
Returns	
—Net return on capital	
employed—before interest	14.2
—after interest	8.2

* Average purchase cost 2.91p/therm or say US¢52.4/MMBtu

supply for the industry when natural gas supplies begin to run out, and in the more immediate future it could become a useful supplement to natural gas, particularly during peak periods of demand.

Recent Legislation

The enactment of the Petroleum and Submarine Pipe-lines Act in 1975 created the British National Oil Corporation, instituted PRT (petroleum revenue tax), and established the main parameters for licences, pipeline activities and participation arrangements for BNOC. However, any contracts for the sale of natural gas concluded before 30 June 1975 were exempt from PRT. This exemption did not apply to any gas liquids which may be extracted from such natural gas before the gas is sold to British Gas.

A year later, on 22 November 1976, the Energy Act 1976 was passed into law. This Act, inter alia, modified and strengthened the existing discretionary controls in the Continential Shelf Act of 1964 over the supply and use of gas (methane, ethane, propane and butane) derived from the UK continental shelf. The Act incorporates provisions which extend the powers of the Secretary of State so that, for example, his consent is now required for the liquefaction of methane and ethane, for the flaring of gas, and for the use of natural gas as a fuel for power-generation purposes.

To what extent this recent legislation is likely to affect the rate of exploration for, and development of, new gas resources remains to be seen. Much will obviously depend on how the government of the day exercises its powers and directs or influences the policies of BNOC and British Gas. There is as yet insufficient evidence to enable anyone to make a confident prediction on such matters. Time alone will tell.

3: The United States

While the first recorded commercial use of manufactured gas was in the United Kingdom, the United States can fairly lay claim to being the birthplace of the natural gas industry as we know it today. There are, of course, many interesting references in literature to the burning of natural gas from underground seepages in China and elsewhere many centuries ago, but these were very small, localised events. As mentioned in Chapter 1, only the Chinese got as far as collecting such gas and piping it over very short distances in hollowed-out bamboo canes. All these early events petered out in time and it was the Americans who made the first serious attempts to utilise natural gas in a practical way.

Whereas in most industrial countries the gas industry was founded and developed over a long period of time solely on the basis of low calorific value gas manufactured from coal, in the United States both manufactured gas and natural gas were developed simultaneously, albeit along quite separate lines. For many years manufactured gas was used primarily for illumination, while natural gas was utilised mainly for industrial applications and heating purposes. Today natural gas comprises almost all of the supplies utilised by the American gas industry, but this was not always the case as we shall see.

Historical Background

The first known use of natural gas in the United States was in 1821 in Fredonia, New York. The accidental ignition by small boys of a seepage of natural gas at the nearby Canadaway Creek brought home to the local townspeople the potential value of this 'burning spring'. They drilled a well 27 feet deep and piped the gas through small hollowed-out logs to several nearby houses for lighting. These primitive log pipes were later replaced by a three-quarter inch lead pipe made by William Hart, the local gunsmith. He ran the gas some 25 feet into an inverted water-filled vat, called a 'gasometer', and from there a line to Abel House, one of the local inns, where the gas was used for illumination.

3.1 Gas service began in Atlanta, Georgia, on Christmas Day 1855 with the ceremonial lighting of the first of 50 gas street lamps (*Atlanta Gas Light Company*)

In December 1825 the *Fredonia Censor* reported: 'We witnessed last evening burning of 66 beautiful gas lights and 150 lights could be supplied by this gasometer. There is now sufficient gas to supply another one [gasometer] as large.' Fredonia's gas supply was acclaimed as: 'unparalleled on the face of the globe'.

This first practical use of natural gas in 1821 was only five years after the birth of the manufactured gas industry in the United States, which most commentators agree was marked by the founding of the Gas Light Company

3.2 Construction of a gasholder for manufactured gas at Thurmond Street, Atlanta, Georgia, 1884 (*Atlanta Gas Light Company*)

of Baltimore in 1816. Let us turn now to manufactured gas and come back later to review how the natural gas business developed.

'The Heavy Charge of the Light Brigade'

Gas lighting based on manufactured gas soon spread from Baltimore to other American cities, mostly for street lights as home gas lighting was initially too expensive for most householders—hence the title of this particular section, coined from an article published in these early days. Many companies were founded in the 1820s and 1830s, including the New York Gas Light Company, which was granted a charter in March 1823 and commenced supplying gas in 1825. The price of gas at that time was $10 per thousand cubic feet. By 1828 New York's famous Broadway was lit by gas lamps.

In 1834, the city of Philadelphia decided to build its own gas works and in so doing became the first municipally owned gas company in the United States. Some other cities where manufactured gas companies were established around this period were Louisville and New Orleans in 1832, Pittsburgh in 1836,

Washington in 1848, Brooklyn in 1849, Chicago in 1850, and Atlanta in 1855. The first issue of *The American Gas Light Journal*, published on 1 July 1859, listed 183 gas light companies in the United States; by 1870 a lighting service based on manufactured gas had been established in 46 American cities. And in the early 1860s gas began to be used for cooking as well as for lighting.

A census published by the *Gas Light Journal* on 1 March 1886, discloses that the number of gas light companies had grown to 971, and that a total of 186 901 public lights were now lit by gas. The average selling price of gas at that time was assessed to be $1.77 per thousand cubic feet, but this masked a very wide price range around the country of from ¢ 75 to $20 per thousand cubic feet. The majority of the companies listed were manufacturing and selling gas made from coal, but several hundred were using oil-gas and other types of gas. Total industry production at that time was reported to be some 23 500 million cubic feet.

Until about 1910 illumination for home and street lighting continued to be the principal use of manufactured gas; other applications such as cooking, water and space heating were still relatively small outlets compared with illumination. Meanwhile, from its tiny beginnings in Fredonia, natural gas was being developed for industrial and heating purposes. The first recorded large-scale—that is to say large scale by the standards of the day—industrial use of natural gas was in the steel and glass works of Pittsburgh in 1883; within a few years natural gas was also being sold in Pittsburgh for domestic purposes.

All this early development of natural gas was confined to outlets which were close to the producing fields as the problems of piping such gas in substantial quantities over long distances had yet to be overcome. As much of the gas came from small, shallow fields, and as gas-producing techniques were still very much in their infancy, an unhappy and chaotic cycle of events developed in the late 1880s and the early part of the twentieth century.

A small gas field would be discovered and a local distribution system would then be quickly developed attracting new industries and other customers to the locality. Reserves were then often rapidly depleted, aided by considerable waste of gas at the point of production, so that, after a few years, curtailment followed subsequently by a total cessation of supplies frequently occurred. Many newly formed gas companies went out of business during these years after a short and hectic business life.

The Wastage of Natural Gas

Elsewhere, particularly in oil fields where there were few, if any, local marketing prospects, enormous volumes of associated gas were flared. Even non-associated gas fields were vented to atmosphere in the vain hope that the non-existent oil lying under the gas would be forced or drawn up to the surface. Gradually legislation was introduced on a state-by-state basis to conserve or prevent this wastage, but even as late as 1934 more than half the gas produced in the Panhandle field in Texas, one of the largest gas fields that has ever been discovered in the world, was still being wasted. Indeed the wastage of natural gas continued to increase over the years and it was not until 1947 that the absolute amounts of natural gas wasted or lost reached a peak of just over 1 trillion cubic feet (Tcf) per year. Few people have been willing to guess how much natural gas was flared or lost in the United States from the early days of oil and gas production to the late 1940s, but one well-known geologist, L. F. Terry, has put the quantity as high as 76 Tcf. The accuracy of this estimate is largely immaterial, but it serves to illustrate the enormous quantities of natural gas that were wasted in earlier years.

The Decline of Manufactured Gas

The use of natural gas primarily for industrial and heating purposes at low prices accounts for the fact that, in volumetric terms, sales of natural gas have exceeded sales of manufactured gas as far back as reliable records have been kept, but it was not until 1935 that sales proceeds from natural gas overtook manufactured gas proceeds. Indeed it is not always appreciated that in spite of an abundance of natural gas, sales of manufactured gas continued to expand up to 1948 and that around this time there were as many as 8.7 million customers. It was only in the late 1940s and in the 1950s that manufactured gas began to be phased out to be replaced ultimately by natural gas. During this transition or conversion from manufactured to natural gas many manufactured gas companies became part of the distribution sector of the natural gas industry. At that time some of these companies distributed both manufactured and natural gases, others an interim blend of both gases, while yet others followed the now established practice of converting from low calorific value manufactured gas straight to high calorific value natural gas.

The Development of Long-Distance Pipeline Systems

A vital factor in this story of the initial parallel but separate development of the natural gas and manufactured gas industries in the United States was the availability of high-tensile steel pipe and the development of techniques to construct, weld and lay large diameter, high-pressure pipelines over long distances. The first long-distance all-welded pipeline was laid by the Magnolia Gas Company of Dallas in 1925. This line from northern Louisiana to Beaumont, Texas, was 217 miles in length and comprised pipe of 14, 16 and 18 inch diameter. Other similar pipelines of several hundred miles each in length were laid by various companies in the ensuing years; in 1931 the first 1000 mile,

3.3 A Buckeye pipeline ditching machine at work near Yazoo City, Mississippi, 1929 (*Southern Natural Gas Company*)

24 inch diameter line was laid by the Natural Gas Pipeline Company of America from Panhandle, Texas, to Chicago.

Continued improvements in pipeline welding and laying techniques, the availability of higher tensile strength steel and larger diameter pipes, hastened the development of a country-wide pipeline network. In 1947 two pipelines known as the 'Big Inch' and the 'Little Inch', which were built during the Second World War to carry petroleum products from east Texas to Pennsylvania, were converted to carry natural gas. These lines helped to open up to natural gas very important markets in the United States east coast area. It is interesting to note that during their initial period of operation these lines transported 140 million cubic feet per day (MMcf/d) without the aid of any compressor stations en route. The subsequent addition of 21 compressor stations increased delivery capacity over threefold to 435 MMcf/d.

By 1955 the total mileage of transmission pipelines amounted to over 145 000, and by 1966 natural gas was available in every one of the Lower 48 states, i.e. in all states south of the US/Canadian border.

To sum up this far, the enormous expansion of the natural gas industry in the United States, at the ultimate expense of the manufactured gas industry, that began to get under way in the late 1920s, but really got into its stride in the 1940s probably owes its success to three main factors: an abundance of gas;

3.4 Four 1250 horsepower compressors with Worthington Harz engines, Reform, Alabama, 1929 (*Southern Natural Gas Company*)

the development of large-diameter, high-pressure pipelines, and attractive (to the end-consumer) prices.

Competition and Its Effect on Gas Usage

Another facet of interest concerning the development of the American gas industry has been the change in the end-use of gas. In the early years illumination was the main domestic use of gas, but the rapid introduction of electricity for lighting just before the turn of the century saw the gradual decline of gas for this purpose. However, the gas industry was quick to react to this unexpected competitor and turned its attention to the cooking and water-heating markets where it in turn increasingly displaced coal. Over the years coal thus lost most of its domestic cooking and water-heating markets to gas. It also lost its market as a feedstock for the manufactured gas industry, but more or less in parallel to this it gained a new market as the main fuel for power generation.

3.5 Completion of a 10 inch pipeline near Columbus, Georgia, 1930 (*Southern Natural Gas Company*)

In the industrial market carbon-black production, for the manufacture of motor-car tyres and as a colouring for printing ink, was an important outlet (some 14 per cent of total sales in 1937) for natural gas. However, to an increasing extent this was seen as a wasteful use of natural gas and this market declined very sharply after 1950 with the majority of carbon-black plants switching over to oil product feedstocks.

The above examples—there are others—illustrate all too clearly how the advancement of one particular form of energy in any particular market or end-use sector is almost invariably wholly or partly at the expense of some other form of energy. A more important point perhaps, which can so easily be overlooked, is that the end-uses of energy have to be flexible to meet changing economic, technical and social needs—such changes are often difficult or impossible to predict with any degree of certainty. The lesson to be learnt from these historical developments is that what is today regarded as a traditional use of a particular form of energy can well be upset violently at some future date for a whole variety of reasons. Apart from supply and other related constraints, the planner or forecaster must be wary of extrapolating past trends—real life can and frequently does turn out to be drastically different.

Pressures Build up for Legislative Controls

High on the list of any review of the American gas business must be the subject of price regulation. It is all too easy today to condemn and to blame price regulation for all the difficulties now being experienced by the American gas industry. While many informed observers consider that price regulation has long since outlived its useful purpose and indeed is now a positive hindrance to the well-being of the industry, there are nevertheless some sound historical reasons for its introduction.

In the 1930s there was much conflict and litigation between, on the one hand, the established manufactured gas industry, which wanted to introduce natural gas supplies gradually so as to prolong the life of its manufacturing assets, and on the other hand, the general public and local civil servants who wanted to utilise natural gas as quickly as possible because of its lower cost and technological advantages. Other accusations against the manufactured gas companies of unfair and monopolistic practices were investigated, inter alia, by the Federal Trade Commission. Undoubtedly there were cases of some companies blocking or hindering the activities of others. The frustration in particular of a group of cities in the Midwest, which were unable to obtain natural gas supplies, led to an alliance of these cities which sought anti-trust action and federal regulation of interstate sales of natural gas.

The Natural Gas Act

These and other pressures led to the Natural Gas Act, which envisioned a regulatory system wherein the Federal Power Commission (FPC) would regulate the interstate pipelines, i.e. in effect transmission companies, while state authorities would continue to regulate intrastate gas distribution companies. The fact that the Natural Gas Act was approved by Congress in June 1938, without a single dissenting vote in either house, demonstrates clearly the support and need for such legislation at that time.

The Natural Gas Act was administered until October 1977 by the permanent staff of the FPC headed by five appointed commissioners. The Act required, inter alia, under its so-called certificate authority, a matching of markets and gas supplies as a requirement for the certification of new sales. It also gave the FPC authority to approve or disapprove natural gas imports and exports and to mandate gas service to communities that had not been able to obtain gas, but which were located near an interstate pipeline. The Act was, of course, designed primarily to protect the interests of consumers. But it had other benefits such as providing a judicial process for settling disputes between competing pipeline companies, between pipeline companies and producers, and also as between pipeline companies and distribution companies.

While the Act envisioned a system of regulation in which state and federal controls would meet without overlap or omission, in practice the FPC consistently maintained over the 16-year period from 1938 to 1954 that it did not have any regulatory authority over independent producers, i.e. producers which were not affiliated with interstate pipeline companies.

The Phillips Decision

In 1950, Congress passed the Kerr Bill, which specifically exempted wellhead gas prices from direct FPC regulation and as such attempted to settle ambiguities in the Natural Gas Act before the Supreme Court had ruled on them. This Bill was vetoed by President Truman because his administration considered regulation of wellhead prices was necessary as otherwise a lack of competition between producers might result in prices increasing beyond reasonable levels.

On 7 June 1954, the Supreme Court in the Phillips Petroleum Company versus Wisconsin Public Service Commission case, in a divided opinion (five to three), interpreted the Act as vesting regulatory authority over producers in the FPC, also that statutory exemption of production and gathering of natural gas was inapplicable to sales of gas to interstate pipeline transmission companies. The so-called Phillips decision was a significant milestone in the history of regulation in that it brought the wellhead price of gas for resale in interstate commerce quite clearly under FPC jurisdiction, but not that gas which was produced for resale in the intrastate market.

One month after this decision the FPC issued its first regulations governing independent natural gas producers, following which many thousands of rate schedules and applications were filed. Within 12 months the FPC had received 10 978 rate filings and 6047 applications for certificates from independent producers. As some cynics have since observed, the regulation of producers became an industry in its own right!

Two years later, in 1956, Congress passed the Harris–Fulbright Act which, like the earlier Kerr Bill, exempted wellhead prices from FPC regulation. However, this Bill was also vetoed, this time by President Eisenhower, who favoured the legislation, but said he could not approve it because of the 'arrogant lobbying' employed on its behalf and because of allegations of producer vote-buying.

Under both the Kerr and Harris 'deregulation Bills' the FPC would still have retained some measure of indirect control over wellhead prices since, under its existing authority to regulate pipeline rates, the FPC could have rejected any excessive gas purchase costs which appeared in the 'cost of service' application of a pipeline company petitioning for higher rates. In the event these bills did not become law and subsequent efforts to change or redefine the responsibilities and authorities of the FPC all came to nought.

The FPC began to apply to wellhead gas pricing the same regulatory procedures used in electric utility and gas pipeline rate-making. However, in practice only 11 full-scale producer rate cases were heard during the 1954–60 period. All except one, which was never concluded, showed producers' revenues to be less than their costs, implying that either oil was subsidising gas production or that the cost data for gas were exaggerated. Within the pricing guideliness established by these proceedings, the FPC accepted 11 091 rate schedules and 33 231 supplements from 3372 producers during this period. Nevertheless by 1960 a very substantial backlog of 3278 filings existed in which rates were suspended pending hearings.

The Area Approach to Price Regulation

The FPC's difficulties in attempting to regulate producers' prices on an individual company cost-of-service basis were summarised by the FPC in the so-called Phillips II opinion. This recognised that the case-by-case system was unworkable and set interim guideline rate levels for various main producing areas. The Supreme Court affirmed this approach in 1963 with the hope that the area approach would prove to be the ultimate solution.

In August 1965 another regulatory milestone was reached when the FPC issued its opinion in the Permian Basin Area rate case, on which hearings had started in 1961. This established uniform ceiling prices applicable to all gas producers in that area. A two-tier system was adopted with a higher price (¢ 16.5 per thousand cubic feet) for new gas to encourage exploration, and a lower price for gas already being produced for sale to the interstate market. Other area rates were subsequently set. Over the period 1954–65, the average national wellhead prices for new gas contracted to the interstate market developed along the lines shown in *Table 3.1*, there being variations, of course, either side of these averages in respect of individual fields or areas as the case may be.

The National Approach to Price Regulation

The hope expressed by the FPC and the Supreme Court in 1963 that the area approach to pricing would simplify matters and facilitate quick decisions was not to be realised. There remained many problems and delays in administering a complex set of different rates for various producing areas and for gases of

TABLE 3.1 AVERAGE WELLHEAD CONTRACT PRICES FOR NEW GAS, INTERSTATE SALES ONLY: 1954–59

Year	*US ¢ per Mcf*	*Year*	*US¢ per Mcf*
1954	11.7	1960	18.2
1955	14.4	1961	17.9
1956	14.8	1962	17.5
1957	16.9	1963	17.0
1958	18.6	1964	16.2
1959	18.4	1965	17.4

Note: For all practical purposes 1 Mcf can be regarded as being equivalent to 1 million Btu.

differing vintages. This led the FPC in June 1974 to abandon the area rate concept in favour of a single nation-wide rate or price. This rate was first set at ¢42 per thousand cubic feet for new gas produced on or after 1 January 1973. In December 1974 the FPC increased the rate to ¢50, with annual 1¢ escalations. On 27 July 1976, the nation-wide rate was again increased to $1.42 for gas from wells commenced on or after 1 January 1975, plus 1¢ per quarter escalation.

All these rates were applicable only to gas sold into the interstate market and were subject to adjustments for calorific content, state severance tax reimbursement, and gas-gathering allowances.

Some Pros and Cons of Regulation

Proponents of the Natural Gas Act would argue with conviction that far from inhibiting the development of the natural gas business, regulation of the interstate pipeline companies restored the respectability of these companies and that, inter alia, it led to a continuous and profitable growth in both sales and monetary terms. There are many who would also argue that the subsequent regulation of wellhead prices in 1954 was a further stabilising influence on the gas industry as a whole.

It is a fact that, over the 30 years or so up to 1970, sales increased nine-fold and natural gas's share of total primary energy consumption in the United States rose from barely 11 per cent to nearly 30 per cent. However, at the same time there existed an abundant supply of natural gas which was consistently sold to end-consumers at prices far below that of electricity, and usually priced marginally below competing oil fuels as well. Undoubtedly these favourable supply and pricing arrangements helped natural gas to achieve its impressive rate of growth. Opponents of regulation would claim that the industry's performance would have certainly been no less impressive without regulation and that regulation, particularly of producers' prices, only stored up problems for the future. But whether exploration efforts and the pricing of gas were aided or hindered by regulation during the period up to the early 1970s must remain a matter of opinion and obviously can never be proved conclusively either way.

One interesting aspect of regulation which deserves highlighting is the impact that the Natural Gas Act had on the business attitudes of the natural gas interstate transmission companies, and perforce their need and desire for growth. The Natural Gas Act enabled the FPC to impose, almost for the first time, public utility type rates of return on transmission companies by the regulation of their activities. Public utility regulation encompasses the need for prudent investment and limits the return (or profit) to a fairly uniform percentage of the depreciated rate base. Under this regime, operating expenses, including the cost of purchased gas, are basically 'pass through' items and result in keeping the delivered cost of gas to a low level. But because the managements of these corporate entities strove quite understandably to increase their companies' annual earnings per share, and hence the dividends payable to shareholders, rate regulations under the Natural Gas Act compelled the transmission companies to expand their capacity in order to achieve increased earnings.

This expansion was apparent to some extent between 1938 and the entry of the United States into the Second World War. After the war existing pipelines were expanded dramatically and many new pipelines were planned and built to take advantage of the plentiful supplies of natural gas that were then available. This compulsive need for expansion, which continued until about 1970, also stimulated a good part of the technological advances in pipeline construction and design that were made over this period.

Whatever the merits of regulation may have been during these years, and undoubtedly there were many, the radical changes that took place in the overall energy scene from the early 1970s called into dispute the whole principle of regulation. The inherent defects and disadvantages of regulation in a basically non-stable situation of world energy prices with the evolvement of new political relationships and strengths, and with the various other social, economic, environmental, and similar pressures that have arisen, became all too apparent.

These matters have been debated earnestly in the United States, and the original concepts and principles of regulation and their inappropriateness to today's circumstances are now recognised and have been changed, not without justification. Certainly few people concerned one way or another with the well-being of the gas industry, be they politicians, governmental agencies, or the industry itself, would deny that changes were necessary; it was the nature, extent and timing of the changes required that was endlessly debated and contested.

The Structure of the American Gas Industry

It will have become apparent from the foregoing that prior to the Natural Gas Act, the natural gas industry had already divided itself into three quite distinct segments or activities. In sequential order the first, of course, is the exploration for and production of natural gas. This phase of activity can include the processing and treatment of natural gas at or close to where it is produced. Treatment is usually necessary in order to provide a gas of appropriate quality that is not only suitable for transmitting over long distances, but also to ensure that the quality of gas distributed is compatible in its general combustion characteristics with gases produced from other fields. These activities comprise the so-called production phase and are undertaken by several thousand small independent producers and by most of the major oil companies as well.

The second phase of activity is concerned primarily with transportation, i.e. the transmission of gas by pipeline either intrastate and/or interstate. A typically small pipeline company in the transmission sector can be quite large in comparison with the many small production companies involved in the production phase. Some of the large interstate pipeline companies, such as Columbia Gas, El Paso Natural Gas, Natural Gas Pipeline Company of America, Texas Eastern, Tenneco, and United Gas Pipeline to name but a few, are very substantial enterprises, in terms of net assets and annual turnover, by any standards. Whereas in the past most transmission companies derived the greater part of their revenue from the transmission of natural gas, to an increasing extent many transmission companies are now branching out into other fields of activity, which apart from exploration and production, are quite often totally unrelated to the natural gas industry. Nevertheless, gas transportation by pipeline is still the main business of most transmission companies.

The growth in the country's transmission system over the period of 1950–77 is illustrated in *Table 3.2*: it will be noted that since the peak year of 1975 there has been some apparent reduction in the length of the total system.

Interstate pipelines are the most heavily regulated segment of the American gas industry and any company wishing to transport or sell gas in interstate commerce must first obtain a 'Certificate of Public Convenience and Necessity'. FPC (or the FERC since 1 October 1977) approval is also required before a pipeline company's tariff can become effective and before it can enlarge or abandon any of its facilities.

Finally, before leaving the transportation sector, it is interesting to note that a number of companies have already or will shortly become involved in a new form of gas transportation, namely the trans-ocean carriage of liquefied natural gas (LNG) by specialised insulated ships. The extent to which this new form of transportation will grow in the coming years is, however, problematic, at least as far as the United States is concerned, where no clear policy has yet

TABLE 3.2 TRANSMISSION SYSTEM AND INSTALLED COMPRESSORS: 1950–77

Year	*Transmission system (miles)*	*Installed compressors (000s horsepower)*
1950	109 400	n.a
1955	145 900	4 350
1960	183 700	6 359
1965	211 300	7 736
1970	252 200	9 692
1975	262 600	12 069
1976	258 200	12 046
1977	255 200	12 105

Note: n.a. = not available

emerged on the amount of imported LNG the authorities will encourage, and more importantly, will consider necessary and approve on acceptable terms.

The third main activity is that of distribution. Apart from such direct sales of gas as may be made by producers or transmission companies to large consumers, it is the distribution company which ultimately sells gas to domestic, commercial, and small industrial end-consumers, as well as to large industrial consumers in many instances as well. There are approximately 1700 distribution companies in the United States of which roughly a third are municipally or publicly owned and two-thirds privately owned. Many distribution companies are also so-called combination utilities, selling not only gas but electricity as well; some indeed also sell other services such as steam.

The majority of present-day distribution companies were originally manufactured gas companies which subsequently converted their systems to natural gas quality. In the early days competition between manufactured gas distribution companies was often intense, but within a comparatively short time it was realised that competition in the same geographic location was uneconomic, physically impractical, and detrimental to the public interest.

Increasingly over time franchises were granted to establish the service area for a given company and the service obligations of the company in that area. This in turn led to a form of regulation exercised initially by municipalities, but later usually by state commissions, and not, as in the case of transmission companies and producers, by a federal authority. While the degree and type of regulation exercised varies, it normally takes the form of ensuring that the service provided is safe and adequate, that the rates charged to customers are reasonable, and that there is no unjust discrimination as to the rates and services that are provided to customers in the designated areas.

It is now the exception rather than the rule for more than one distribution company to operate in any one large town or city. In such cases where a city may be served by more than one distribution company, the designated service areas for each such company do not overlap with each other. There is, therefore, no longer any direct competition as between one gas distribution company and another, but, of course, this in no way protects distribution companies from the competitive pressures of electricity, coal, and liquid petroleum fuels.

Unfortunately, although there is a wealth of historical statistical data on the American gas industry, such data do not separate sales made by distribution companies from the totality of all sales which include sales made directly by transmission companies and producers to end-consumers. Statistics on distribution mains, which have increased from over 300 000 miles in 1955 to nearly 670 000 miles in 1977, whilst indicative of how the distribution system has grown, do not reveal the sizes (i.e. diameter) of distribution mains included or excluded from such figures.

This brief description of the three main groupings or segments of the gas industry—production, transmission and distribution—represent the classical segregation of the gas industry's principal activities and are reflected, although not always so distinctly in the corporate sense, in many other countries. The Natural Gas Act reinforced and underlined in a legalistic or regulatory manner the corporate segregation of these principal activities.

The Domestic or Residential Market

As already described, the domestic market, or the residential market as it is usually called in America, was built up in the early days on the use of manufactured gas for illumination. As in so many other countries the invention of the electric dynamo and the development of electricity for general lighting purposes, which in the United States started in the second half of the nineteenth century, was a potential threat to the continued existence of manufactured gas distribution companies; this became a very real and serious threat with the introduction of the electric incandescent lamp a few years later. Although the gas industry was unprepared for this challenge, it fought back and in effect saved itself from financial disaster at that time by the rapid introduction of the incandescent gas mantle. This was invented by the Austrian, Welsbach, and first demonstrated in London around 1877; the American distribution companies were quick to employ this new invention in defence of their business.

The incandescent mantle increased the illumination value of gas lights almost sevenfold from the equivalent of about 3 to 20 candles per cubic foot of gas. With these new mantles, and the inherent advantages of having existing gas service connections to many homes and of gas being cheaper than electricity, the use of gas for illumination increased rather than declined,

3.6 An emergency customer servicing wagon, Brooklyn, *c.* 1900 (*Brooklyn Union Gas Company*)

reaching its peak probably a few years before the First World War. However, thereafter it increasingly lost ground to the admittedly more efficient, cleaner electric lighting; by about 1920 few homes continued to be lit by gas.

While this struggle was going on the gas industry was slowly building up a new important market for gas: this was cooking. Although the gas cooking stove was first demonstrated in the United States as early as 1851, it was not until the early years of the twentieth century that the cooking stove began to gain wide acceptance. Other domestic uses for gas, such as space heating, and then subsequently (around 1926) gas-fired refrigerators, followed. As these new uses for gas developed, they helped initially to offset the decline in gas lighting and, in due course, more than replaced it in terms of gas usage. However, it was not until manufactured gas was displaced by natural gas that gas-fired central-heating systems, air-conditioning units (in 1937) and later dishwashers, incinerators and clothes dryers, became important and growing outlets for gas. Paradoxically in more recent times the wheel has turned a full circle, and with the development of decorative outdoor natural gas lights for gardens, patios, driveways, residential streets and so on, more gas is now consumed for such purposes than was the case at the height of the gas light era 50 years or so ago.

Turning now to some statistics, *Table 3.3* shows how the domestic market has expanded over the last two decades.

Two points are worth mentioning. First, the decline in the average annual

3.7 A dispatching room, Brooklyn, through which customers' requests for service are routed. The map shows the company's gas distribution system as it was in 1913 (*Brooklyn Union Gas Company*)

TABLE 3.3 DOMESTIC CUSTOMERS AND CONSUMPTION RATES: 1945–77

Year	*No of customers (millions)*	*Total consumption (trillion Btu)*	*Average consumption (millions Btu/customer)*
1945	18.6	775	41.6
1950	22.1	1 384	62.6
1955	26.3	2 239	85.2
1960	30.4	3 188	104.8
1965	34.3	3 999	116.5
1970	38.1	4 924	129.2
1975	40.9	4 991	121.9
1977	41.7	4 946	118.7

consumption per domestic customer during the last couple of years or so, after making due allowances for seasonal variations from one year to another, is thought to be the result of conservation practices. Second, although natural gas enjoys a significant price advantage over competitive fuels, and in spite of the increase in new housing starts (e.g. 1976 was up nearly one-third on 1975), the rate of acquisition of new domestic customers has been constrained below the assessed potential for at least the last two years and quite probably rather longer. The basic reason for this is the reduction in new supplies of natural gas, which has precluded many distribution companies from competing effectively for new customers.

For some years now natural gas has supplied more than half the total amount of energy consumed in the domestic sector. Based on data representative of some 28 million domestic customers, i.e. about 70 per cent of the total number of 41.7 million domestic gas customers in 1977, the average consumption rates show in *Table 3.4* were compiled in respect of 1975.

TABLE 3.4 AVERAGE ANNUAL CONSUMPTION PER TYPE OF APPLIANCE/END USE IN 1975

Application/appliance	*Therms**
Cooking stoves (family size)	105
Water heaters	325
Clothes dryers (gas pilot)	73
Clothes dryers (electric pilot)	48
Incinerators	151
Gas lights	184
Air conditioners (consumption per ton)	273
Space heating (all types)	1 069

* 1 therm equals 100 000 Btu

The importance of the space-heating load is self-evident from the above. This is further underlined by the fact that in 1977 over 85 per cent of the domestic gas customers in the United States used gas for space heating; in 1955 and 1965 the comparable figures were 56 and 76 per cent respectively.

Out of the total current American stock of occupied housing units of some 74 million, it is estimated that about 35 million are heated by gas, 10 million by electricity, 16 million by oil, and the balance either by coal, wood, LPG or are not heated at all. These statistics, including those given in *Table 3.4*, should be interpreted with caution as, of course, the actual rates of consumption by type of appliance and the degree of heating penetration achieved vary from one part of the country to another, depending on the climatic, social, economic, and other conditions prevailing therein. Subject either to any significant changes in these latter factors and/or the development of new domestic applications for gas, it is unlikely that any marked increase in average consumption rates will be achieved. Indeed, as already remarked upon earlier, the effect of conservation practices will probably result in some further declines in average consumption rates in the coming years. Future expansion of the domestic market must, therefore, depend primarily on the acquisition of new customers, and this in turn depends on the availability and cost of new gas supplies.

Right from the first days the domestic market has been the bedrock of the distribution companies' gas business; it still is today and is likely to remain so in the foreseeable future.

The Commercial Market

As the name implies the commercial market comprises the use of gas for offices, schools, hospitals, shops, hotels, restaurants, and similar outlets of a non-industrial nature. Whilst manufactured gas at its height enjoyed for some years a significant proportion of the commercial market, natural gas at that time increasingly supplanted manufactured gas for larger volume applications. However, reliable statistics on the historic split of the market between manufactured and natural gas are not readily available for the period when both types of gas were marketed. Furthermore, such statistics as are now available on sales of all types of gas into the commercial market may be somewhat understated, given the mixture of commercial and industrial-type applications for which gas is used by certain individual end-consumers within their own premises.

Table 3.5, which shows how the commercial market has developed since the mid-1940s, relates, therefore, to those outlets which fall clearly in the commercial market classification. Apart from the earlier years, which include an unidentified amount of manufactured gas, it is reasonable to assume that most of the gas sold for commercial applications during this period was natural gas.

To put the above figures in perspective, in 1945 the commercial market represented in volumetric and gross revenue terms some 10 per cent and 13 per cent respectively of total gas sold for all purposes by American transmission and distribution utilities. For 1977 the corresponding figures are 17 and 18 per cent, which illustrate the growing importance of the commercial market in more recent years. For gas utilities, this growth has been enhanced by the improvement in average unit retail prices which have climbed from about ¢60 in 1945 to $2.07 per million Btu in 1977.

TABLE 3.5 COMMERCIAL CUSTOMERS, SALES, CONSUMPTION AND REVENUE: 1945 77

Year	No. of customers (000s)	Total sales (trillion Btu)	Average consumption per customer (million Btu)	Gross revenue ($million)
1945	1 278	250	196	149
1950	1 739	410	236	266
1955	2 048	603	295	424
1960	2 458	920	374	723
1965	2 790	1 345	488	1 054
1970	3 131	2 007	641	1 620
1975	3 367	2 387	709	3 302
1977	3 371	2 409	715	4 980

The Industrial Market

Almost from the beginning the industrial market has been the principal outlet for natural gas in the United States, a situation which still holds good today. Manufactured gas, with its early predominant role as a means of illumination, its subsequent expansion into the domestic, cooking and heating market, and to some extent into the commercial market as well, never made any significant penetration into the American industrial market even in its heyday. For example, of the total sales of manufactured gas in 1932 amounting to just over 2000 million therms, only 8 per cent was sold for industrial purposes.

For many years prior to the advent of long-distance pipelines—one of the factors which eventually led to the demise of manufactured gas—natural gas producers concentrated on securing industrial-type outlets close to the point of production, i.e. within the limits of the technology available at that time. This brought them into fierce competition with coal producers. In contradistinction, the manufactured gas industry, with its emphasis on illumination and other non-industrial uses, was a major customer of the coal industry and enjoyed good relations with it.

With the development of long-distance pipeline transmission systems, the initial customers for natural gas were distribution companies which in many cases had few, if any, large-volume commercial and industrial outlets. This resulted in poor load factors for the gas producers and transmission companies because of the very seasonal nature of the domestic heating market. For the sake of clarification load factor in this context can be defined as the relationship between the average day's delivery of gas and the maximum day's delivery of gas over a stipulated period, say, of one year. This relationship is frequently expressed as a percentage, or in more sophisticated circumstances as so many hours out of 8760 hours, i.e. the number of hours in a year.

In order to improve load factors, and hence to reduce the incidence of very high fixed costs, which exist regardless of the amount of gas produced and transported, pressure was put on distribution companies to develop large volume commercial and industrial outlets. In general, such outlets were not only less susceptible to seasonal fluctuations than the domestic market, but also offered higher throughputs and hence lower unit costs. In addition, some very large off-take customers were supplied direct by both gas producers and transmission companies with the same objectives of improving load factors and increasing sales volumes.

The inherent qualities and characteristics of natural gas such as its clean burning, versatility, ease of control and flexibility in use, the fact that no fuel storage is required on customers' premises, etc., all helped to ensure its rapid penetration of the industrial market. Price, compared with alternative fuels, was also frequently in natural gas's favour especially in more recent years.

Given the predominance of natural gas and the minor contribution by manufactured gas in the industrial market, *Table 3.6* showing the development of this market over the period 1945–77 can be regarded as comprising mostly natural gas; some relatively minor volumes of manufactured gas are, however, included in the first few years of the period concerned.

TABLE 3.6 INDUSTRIAL CUSTOMERS, SALES, CONSUMPTION AND REVENUE: 1945–77

Year	*No. of customers (000s)*	*Total sales (trillion Btu)*	*Average consumption per customer (million Btu)*	*Gross revenue ($million)*
1945	80	1 452	18 150	281
1950	99	2 287	23 189	480
1955	121	3 535	29 168	938
1960	141	4 709	33 495	1 563
1965	166	6 147	36 982	2 148
1970	199	8 439	42 387	3 181
1975	184	6 837	37 259	6 718
1977	173	6 710	38 723	11 385

The above data relate to transmission and distribution companies, which are collectively designated in most American statistics as the gas utility industry. Not included are sales made directly to end-consumers by gas producers, the outlet for which (excluding producers' and their affiliates' own fuel requirements) was mainly for power generation. Certain sales by gas utilities to electric utilities are also excluded. However, none of these statistical reporting complexities distort unduly the general pattern of use and the size of the industrial market as illustrated in *Table 3.6*.

What *Table 3.6* does not reveal too clearly is that the number of industrial consumers has declined since the peak of 209 000 reached in 1972 and 1973, and that industrial sales volumes have also declined from their peak of 8776 trillion Btu in 1972. The main reasons for this are shortages or curtailments in gas supplies, resulting in some industrial gas customers switching to other fuels, coupled with the effect of conservation activities that have been undertaken by many existing customers. However, in spite of the fall-off in sales since 1972, yearly gross revenues have continued to increase as a result of higher unit prices which from a fairly constant yearly average of between ¢30 to 40 per million Btu during the 1950s and 1960s, and ¢40 to 50 in the early 1970s, rose to almost $1 in 1975, and to $1.70 per million Btu in 1977.

Since the early 1950s natural gas has been the predominant fuel for industrial purposes in the United States. By 1962, natural gas's contribution to total primary energy supplies for industrial uses had risen to 39 per cent compared with 27 per cent each for coal and oil, with electricity making up the balance. By 1976, natural gas's share of the industrial sector had increased to nearly 46 per cent and oil's to about 34 per cent, both at the expense of coal, which by then had dropped to just over 20 per cent.

Before concluding this brief account of the industrial gas market it is necessary to draw the reader's attention to three points. First, that no distinction has been drawn between gas used for specialised process uses and that used for under-boiler or steam-raising purposes. There are no reliable nation-wide data which split down the use of gas over these two broadly defined and separate categories, yet in practice, especially in a situation in which natural gas supplies will no longer be abundant, this is or will become an important consideration. Basically process-type applications for gas, in which the inherent qualities of gas have advantages over most other competing fuels, can be regarded as premium uses and as such can or should command better prices than non-premium uses and should have first call on available supplies for the industrial market. For non-premium uses, e.g. under-boiler and steam-raising outlets, other fuels can be used with more or less equivalent performance, provided there are no insurmountable constraints such as the provision of storage facilities (for oil or coal) on the customers' premises, prohibitive legislation on the permitted level of emissions of the products of combustion and so on. Various attempts have been made to split the industrial gas market over premium and non-premium outlets, but the resultant data are of doubtful accuracy and worth.

Second, although there is wealth of published statistics on the American gas industry, the sheer size and complexity of the business involving many thousands of producers, transmission, and distribution companies, makes it virtually impossible to categorise the use of gas over the many types of applications, including electricity generation, that it has been used for within the all-embracing industrial sector in past years. Here again such analyses that have been made tend to be suspect. Accordingly, no attempt is made here to try to differentiate between premium and non-premium uses, or to list the use of gas for specific applications on an historical basis. Considerable efforts are now being made to break down gas usage in this way, but these efforts are directed mainly at the present situation and planning for the future, rather than attempting to analyse the historic situation over the time scale that this book endeavours to encompass.

The third point, which deserves a separate section to itself, is how the American gas industry has developed the concept of interruptible sales—as distinct from firm or continuing sales—mainly for the industrial and power generation markets, which has spilled over to some extent to the commercial market.

Interruptible Sales

As the name implies interruptible sales are those sales of gas which are or can be interrupted for certain specified periods. Obviously where gas is used for a specialised process use it is not practicable for the end-consumer to switch from one fuel to another at short notice. However, there are many cases, in particular for under-boiler firing, where provided alternative burners and related equipment and facilities exist or can be installed at relatively low cost and with little inconvenience, it is possible for the end-consumer to use a variety of fuels.

In order to improve the load factor of gas producers and gas utilities, a highly sophisticated interruptible market has been built up over the years. There is no record of where and who first introduced this concept, but in essence it has developed over time along the following lines.

Interruptible service offers gas producers, transmission and distribution companies the ability to supply certain types of customer with gas when other customers' demands are at or below average yearly off-take levels, and in addition, to cut off gas supplies in favour of firm and higher priority users when their off-take rises above average yearly requirements. Typically this would arise in a cold winter when the need for space heating results in a sharp rise in domestic and commercial consumption. In warmer parts of the country the reverse situation could apply when the need for air-conditioning in the height of summer can exceed normal gas requirements during a mild winter. In other words the interruptible business is a reflection very largely of temperature sensitivity.

An interruptible end-consumer will by definition agree to receive gas supplies on a firm basis for say six to nine months of the year and be prepared, subject to adequate notice, to be curtailed or to forego completely gas supplies during the remainder of the year if the supplier deems it necessary in order that the requirements of higher-priority customers can be met.

In return for agreeing to being interrupted during stipulated periods of the year, and to compensate for the extra cost of installing dual-fired boilers and for storage of alternative fuels, the customer will normally pay rather less for his gas supplies compared with a year-round firm supply customer, other things being equal. The period of interruption, usually related to changes in ambient temperatures, varies from customer to customer, as also does the amount of advance notice the customer will receive before he is interrupted. When interruption becomes necessary, the gas utility will simply shut off its gas supplies and the customer will immediately switch over to his standby alternative fuel until such time as gas supplies are resumed. The latter may be within a day or two, or as long as several weeks, depending on the contractual arrangements and prevailing weather conditions.

With the appropriate blend of firm and interruptible customers, backed up by strategically located underground storage, peak-shaving facilities, and 'line pack' (i.e. the inherent storage capacity of a large pipeline system), some gas utilities have achieved year-round load factors close to 100 per cent during normal seasonal years.

There are no statistics available prior to 1971 giving a break-down between firm and interruptible sales, and even those for subsequent years are not particularly meaningful without reference to the ambient temperatures that prevailed over the actual period when interruption applied. Suffice it to say that for electricity generation, the market most amenable to the interruptible concept, interruptible sales have usually been substantially greater in volumetric terms than firm sales year-on-year over the period 1971–77.

Although there are critics of the use of natural gas for under-boiler and other non-premium applications for which coal and high sulphur content fuel oil can be used with more or less equal facility, nevertheless the ability to use gas on an interruptible basis for such outlets has undoubtedly helped to keep unit costs, and hence prices, down for all classes of customers.

The 1973 Shocks

This was a traumatic year in several respects. The most obvious shock to the world at large was the decision by the Organisation of Petroleum Exporting Countries (OPEC) to more than double the price of internationally traded crude oil. Further price rises followed during 1974 and 1975. As shown in *Table 3.7* the increases were indeed dramatic and unparalleled in history.

By comparison the average wellhead price of indigenous natural gas in the United States over the same period rose from less than ¢20 to just under ¢45 per million Btu. Direct comparison of international crude oil prices with indigenously produced natural gas is not, of course, entirely valid: the former

TABLE 3.7 SAUDI ARABIAN LIGHT: OPEC MARKER CRUDE OIL PRICES

Effective date	*$ per barrel*	*¢ per MMBtu*
1.1.73	2.05	35
16.10.73	4.76	82
1.1.74	10.95	188
1.6.74	11.05	190
1.10.74	10.84	186
1.11.74	10.46	179
1.10.75	11.51	197

being a relatively small part of America's total primary energy requirements at that time, whilst the latter, apart from some Canadian gas imports, represent almost the totality of the country's gas supplies. The subsequent impact that these OPEC decisions had on the world's economy and on overall energy prices and values is now too well known to require yet another detailed account here. Quite simply it meant that controlled or regulated natural gas prices in the United States lagged increasingly behind the prices of competing fuels, notably, but not exclusively, oil products. The latter were driven upwards inexorably by the twin motors of world-wide inflation and the rapid price rise in internationally traded oil, increasing quantities of which the United States was obliged to import as its own crude oil production became insufficient to meet domestic needs.

The situation this brought about was recognised by President Nixon, but his 'Project Independence' was, in retrospect, impractical and too ambitious. The next major attempt by President Carter to reorganise and change past trends in American energy usage and pricing, the National Energy Plan, is, however, discussed later.

The second shock of 1973 was that this was the first year in which indigenous natural gas production, and hence consumption, peaked out after an almost continuous annual expansion since the middle of the nineteenth century. Perhaps this was less of a surprise to some of those more intimately involved in the natural gas business as in 1968 net production for the first time exceeded annual additions to proved reserves. Since then new reserves discovered in the Lower 48 states have never exceeded or equalled annual rates of production. There was, however, a sharp one-time uplift in the total US reserves portfolio with the discovery of Alaskan North Slope reserves in 1970, but even these reserves, important though they are, did not fundamentally affect the continuous, progressive annual decline in the reserves: production ratio. For the US as a whole this ratio dropped from a healthy 32.6 in 1946 (the first year for which statistics are available), to 15.9 in 1967, to a worrying low of 10.7 in 1977, or 9.2 if Alaskan reserves are omitted. For those readers less familiar with industry terminology, this meant that in 1977 total proven reserves, which by definition exclude future expectations of new discoveries, were only sufficient to last a further 10.7 years at current rates of production and consumption.

TABLE 3.8 UNITED STATES TOTAL NATURAL GAS SUPPLY AND PRODUCTION: 1946–77 (IN TRILLIONS CUBIC FEET)

Year	*Reserves added*	*Net annual production*	*Year-end proven reserves*	*R/P ratio*
1946	17.6	4.9	160.4	32.6
1950	12.0	6.9	185.3	27.0
1955	21.9	10.1	223.2	22.2
1960	13.9	13.0	263.0	20.2
1965	21.3	16.3	287.1	17.6
1967	21.8	18.4	293.7	15.9
1968	13.7	19.4	287.8	14.3
1970	37.2	22.0	290.7	13.2
1973	6.8	22.6	250.0	11.1
1977	11.9	19.4	208.9	10.7

Note: Apparent discrepancies in the above table are the result primarily of net additions to or withdrawals of gas from underground storage.

While *Table 3.8* portrays the overall situation, the reserves position for the interstate and intrastate gas markets was rather different. In the case of the interstate market, reserves (excluding Alaskan gas which has yet to become available) declined from 198 Tcf in 1967 to 96 Tcf in 1977. But production from such reserves only dropped from 10.5 Tcf to 9.4 Tcf over this period resulting in reserves: production ratios of 18.9 in 1967 and 10.2 in 1977. The situation for the intrastate market was that reserves declined from 95 Tcf in 1967 to some 81 Tcf in 1977, with annual production increasing from 7.9 Tcf in 1967, to 10.6 Tcf in 1973, falling somewhat to an average of around 9.8 Tcf per annum over the years 1975, 1976 and 1977. For the intrastate market the reserve: production ratios for 1967 and 1977 were, therefore, about 12.0 and 8.3 respectively.

The swing of gas production and disposal from the interstate to the intrastate market since the late 1960s/early 1970s reflected the greater attractiveness to producers of the intrastate market, where wellhead prices were not subject to Federal regulation. But irrespective of these marketing changes the overall trend was clear—since 1967 the United States has been living today on tomorrow's gas reserves.

Much has now been written by many organisations and individuals on how they foresaw this impending gas supply shortage and its likely consequential effects. Be that as it may, in fact those who actually stated clearly at the time what was likely to happen were relatively few and far between and their concerns went largely unheeded for several years. Indeed even after the peak year of production (1973), some doubts were still voiced that the gas industry was simply crying wolf with the prime objective of trying to raise prices, whereas responsible industry spokesmen were much more concerned about the future viability of their business: unless the trend in declining supply availability was arrested and hopefully reversed, the long-term outlook looked bleak.

The Emergency Gas Act of 1977

The unexpected severe gas curtailments of the winter of 1976/77 brought home to everybody the critical situation the gas industry was now facing. Shortly after assuming office, President Carter addressed himself to this immediate problem and his proposed legislation—the Emergency Natural Gas Act of 1977—was passed quickly by Congress on 2 February 1977, with virtually no amendments and with little opposition. The Act contained two major provisions permitting Federal allocation of gas among interstate pipelines until 30 April 1977 and purchases by interstate pipelines and industrial customers from both producers and intrastate pipelines at prices which 'the President determines to be appropriate'. In practice under this law the FPC only authorised emergency sales of less than 0.1 Tcf at prices averaging $2.25 per thousand cubic feet.

However, the industry learnt much from the problems of that winter and the less severe winter of 1977/78 caused no major unscheduled curtailments. Indeed, although the volumes of produced gas were still declining year-on-year, supply flexibility and system capacity was corresponding enhanced, putting the industry in a somewhat better position to cope with sudden changes in demand.

The Demise of the FPC

On 10 June 1920, President Wilson signed into law legislation creating the Federal Power Commission to provide an orderly means of developing the nation's hydro-electric resources. As mentioned earlier, in June 1938, Congress enacted the Natural Gas Act which gave the FPC jurisdiction over the transportation of natural gas in interstate commerce, the sale of natural gas in interstate commerce for resale for ultimate public consumption, and regulation of natural gas companies which transport or sell interstate.

Whilst the debate described in the next section on the National Energy Plan was continuing during 1977 and 1978, President Carter's administration took steps to reorganise the many Federal regulatory bodies and energy agencies which had come into being at various times over the previous 50-odd years. The prime objective of this reorganisation was to establish within the Federal government one principal executive branch to ensure effective management of a co-ordinated national energy policy. These proposals met with relatively little opposition and on 4 August 1977, the Department of Energy Organisation Act became law. For natural gas this meant the abolishment of the FPC on 30 September 1977 and the creation, as an independent commission within the Department of Energy (DOE), of the Federal Energy Regulatory Commission (FERC) on 1 October 1977. Simultaneously the FERC took over most, but not all, of the functions (and staff) of the former FPC.

A rose by any other name some might say, but this was not quite the case. Among other things another body within the DOE, the Economic Regulatory Administration (ERA), was accorded a related but separate role on gas affairs with particular regard to gas import applications. After some initial confusion a broad understanding was reached between the FERC and the ERA on their respective roles in such matters, although some further clarification on certain detailed points has yet to be announced.

In essence, the ERA is concerned with determining whether import proposals are in the national interest, their effect on the US balance of payments, security of supply, pricing aspects at the point of importation, and certain other matters. The FERC for its part has responsibility to approve the siting, construction and operation of particular facilities, and the place of entry for an import. Furthermore, if an importer proposes to sell natural gas in interstate commerce for resale, or to have imported gas transported by interstate pipeline companies, the FERC will review the applicable prices and arrangements under the appropriate standards of the Natural Gas Act. The FERC is also, of course, the principal regulatory body for matters of internal jurisdiction relating to indigenous gas sales, etc., previously carried out by the FPC.

In practice, if the ERA administrator determines within his delegated jurisdiction that a proposed import is not consistent with the public interest, then subject to any rehearings and subsequent reviews, the further considerations of the FERC would not be undertaken. However, it should be made clear that as the law now stands, neither the ERA nor the FERC have exclusive authority over the importation, transportation, or sale in interstate commerce of natural gas. But if one of these two bodies denies an application, then the other would not normally proceed with its deliberations. Further complications are that the ERA in approving an import can impose conditions requiring that the imported gas be marketed incrementally through separate contracts. In the event of such a conditional authorisation by the ERA, the FERC has the power to reject the import in toto on the grounds that there is not a market for the gas on an incrementally priced basis in the United States, and, therefore, the import is not economically feasible. On the other hand, the FERC has no power to overrule the ERA on whether the gas should be marketed incrementally or on the basis of rolled-in pricing.

The National Energy Plan

One of the first problems President Carter turned his attention to after being elected to office was—to use his own words—the US energy crisis. This led to the publication on 29 April 1977 of his 'National Energy Plan', the salient features of which were:

—conservation and fuel efficiency;
—rational pricing and production policies;
—reasonable certainty and stability in government policies;
—substitution of abundant energy resources for those in short supply; and,
—development of non-conventional technologies for the future.

One of the several main elements of the plan was to ensure that: 'The pricing of oil and natural gas should reflect the economic fact that the true value of a depleting resource is the cost of replacing it.' For gas it was stated that: 'the artifically low price of natural gas has encouraged its use by industry and electric utilities, which could use coal, and in many areas has made gas unavailable for new households, which could make better use of its premium qualities.'

For natural gas the plan proposed, inter alia, that all new gas sold anywhere in the country from new reservoirs would be subject to a price limitation of approximately ¢175 per thousand cubic feet at the beginning of 1978; the interstate–intrastate distinction would disappear for new gas; higher incentive pricing for specific categories of high-cost gas (e.g. from deep reservoirs, tight formations, etc.); the cost of more expensive new gas would be allocated initially to industrial rather than commercial or residential users; prices of existing supplies would be guaranteed with adjustments to reflect domestic price increases; production in excess of contracted volumes from existing reservoirs would qualify for a price no higher than the then current ceiling of ¢142 per thousand cubic feet. For LNG imports the plan proposed examination of projects on a case-by-case basis, with the previous Energy Resources Council guidelines being replaced with a more flexible policy that sets no upper limit on LNG imports. However, attention would be paid in each import application on the reliability of the selling country, the costs and safety conditions involved, etc.

Many aspects of the plan were debated in great detail and in considerable depth, both dispassionately and with varying degrees of emotion, by the public, the industry, politicians, the media, and indeed almost everybody during the remainder of 1977 and throughout most of 1978 as well. Perhaps the issue that attracted the greatest attention and emotion was that of natural gas pricing. Opinions ranged from immediate deregulation of all gas prices, through all manner of combinations and variations, to comprehensive and continued regulation for an indefinite period. Many proposals were drafted only to be rejected or defeated at some stage or the other before they could be enacted into law.

By the late summer of 1978 it was apparent to most lobbyists and politicians that if any changes to the present pricing system were to be made—and most politicians were by then convinced that some changes were necessary, even if they were still not agreed on what form such changes should take—some compromise package of pricing proposals was necessary. The alternative, to leave matters as they were, was generally unacceptable, but to procrastinate further would perhaps defer the issue for several years, given the precedent of past abortive attempts under Presidents Truman and Eisenhower.

In the end after some hectic negotiations between various leading politicians and interventions by the White House, the Senate finally passed by 57 votes to 42 the so-called Natural Gas Conference Bill on 27 September 1978 and this was ratified by the House of Representatives on 15 October 1978 by 231 votes to 168. This Bill, which then became the Natural Gas Policy Act, was enacted into law on 9 November 1978 by President Carter with an effective date of 1 December 1978. After nearly 18 months of earnest debate and countless proposals, agreement at political level had at last been reached.

The Natural Gas Policy Act

As will have been gathered from the previous section the Natural Gas Policy Act (NGPA) was very much a compromise proposal which pleased few in its totality, but at least was regarded as being some improvement on the past system by most. Instant price deregulation was not to be, to the disappointment of those supporters of deregulation who considered, and presumably still do, that gas prices should find their own market level so as to allow natural gas to compete with alternative energies on a proper market value basis and, more importantly perhaps, to stimulate exploration activity. Equally the easement of ceiling prices with escalation provisions, particularly for so-called 'new gas', and the bringing under Federal control of intrastate gas sales with appropriate safeguards, both for existing interstate and intrastate on-going contracts, was a step in the right direction.

How effective the NGPA will be in evening out the disparity of supply availability as between the intrastate and interstate markets, encouraging exploration for new gas reserves in difficult, high-cost locations, and in reducing, over time, the disparity between gas prices and other energy prices, remains to be seen. Such judgements can only be made retroactively by historians in years to come.

At this time it is not yet possible to analyse the full consequences of the NGPA provisions. The provisions of the Act are extremely detailed and complex. It provides for in excess of 22 different ex-field prices and the ground rules for categorising which gas falls under which price regime are now being formulated by the FERC. It will be some time before these ground rules are established in such a form as no ambiguity exists. Meantime, the FERC has published its interpretations of the provisions contained in the NGPA and indicated the maximum lawful prices that are expected to apply in January 1979. *Table 3.9* illustrates some of these prices but does not attempt to cover each and every pricing permutation.

TABLE 3.9 ASSESSED MAXIMUM LAWFUL PRICES: JANUARY 1979 (US ¢ PER MILLION BTU)

Gas not committed to interstate commerce on 8.11.78	*¢ per MMBtu*
New natural gas and natural gas from a new Outer Continental Shelf (OCS) reservoir on an old OCS lease which was not discovered before 27.7.76	209.6
New onshore production wells not within 2.5 miles of any 'marker' well and/or is at least 1 000 feet deeper than the deepest completed marker well within a 2.5 mile radius	198.0
High-cost gas from wells deeper than 15 000 feet for which drilling began on or after 19.2.77	209.6
'Stripper' well natural gas	224.3
Natural gas not otherwise covered	163.9
Certain gas committed or dedicated to interstate commerce on 8.11.78	
Post-1974–all producers	163.9
1973–74 biennium gas—small producer	138.7
—large producer	106.4
Permian Basin gas—small producer	46.5
—large producer	40.7
Appalachian Basin gas, large producers, contracts dated prior to 8.10.69	34.6

With effect from 1 January 1985, price controls will be lifted from 'new natural gas', new onshore production wells deeper than 5000 feet, and intrastate gas selling for more than ¢100 per million Btu. Certain high-cost gas will be deregulated from a date to be established by the FERC. Gas from new onshore wells shallower than 5000 feet will be deregulated from 1 July 1987.

However, the President or Congress may reimpose price controls for one 18-month period; such reimposition may not take effect earlier than 1 July 1985, nor later than 30 June 1987. All other gas will remain under price controls indefinitely.

Under the NGPA the FERC's jurisdiction encompasses both interstate and intrastate sales; the latter hitherto were not subject to Federal price regulation. Furthermore, the NGPA made it unlawful for new sales to be sold in excess of the statutory maxima which are ceiling prices. However, the latter do not supersede existing contractual agreements fixed at lower prices. For intrastate gas the principal maximum lawful price provisions are:

—if the contract price applicable on 9 November 1978 was not more than ¢206 per million Btu, the maximum lawful prices shall be the lower of the monthly existing contractual price, or the maximum lawful monthly price for new gas;
—if the contract price applicable on 9 November 1978 was greater than ¢206 per million Btu, the maximum lawful monthly price shall be the higher of the maximum lawful monthly price for new gas, or the contract monthly price adjusted for inflation.

All the above representative selection of prices, and the many others omitted, are subject to prescribed monthly rates of indexation. These are based upon the gross national product (GNP) implicit price deflator. This deflator takes into account price changes in all goods and services in the economy and is adjusted by the addition of a Consumer Price Index 'correction factor'. In addition, for certain categories of gas, mainly but not exclusively new gas and high-cost gas, the price is also escalated by a 'real growth factor'. This latter adjustment is 3.5 per cent for any month before April 1981 and 4.0 per cent for any month after May 1981.

If the reader is confused by the complexity of prices introduced by the NGPA, then he may take heart that those who have to apply such prices, be they producers, gas utilities, or administrators, are probably no less confused as to their practical application where any particular gas supply does not fall precisely into any one particular category.

Apart from wellhead pricing the NGPA also addressed certain other issues such as incremental pricing, Presidential authorities as to the purchase and allocation of natural gas supplies under emergency conditions, and policies prohibiting the curtailment of supplies to specified categories of customers. The NGPA did not, however, deal with LNG imports, various safety and environmental issues, etc., for which policy statements and/or legislation were promised but have yet to come. On the matter of LNG imports in fact the intention of the Carter administration is now not to issue any general policy statement, but to let the outlines of an LNG policy emerge as a consequence of specific decisions on individual import applications.

Exports and Imports

The United States has exported natural gas by pipeline to Canada and Mexico for well over 30 years, and LNG by ship to Japan since 1969. Collectively these exports reached a peak of some 93 700 million cubic feet in 1968, equivalent to nearly 260 million cubic feet of gas per day. This was less than 0.5 per cent of total indigenous production for that year. By 1977, total exports had declined to 55 600 million cubic feet of which nearly 52 000 million cubic feet comprised LNG to Japan.

Imports of natural gas by pipeline from Canada commenced in 1951 and since 1971 have run at a fairly constant annual rate of between 2500 and 2800 million cubic feet per day. Pipeline gas imports from Mexico commenced in 1955, but after reaching an average rate of just over 100 million cubic feet per day in the late 1960s, they fell away quite sharply in the early 1970s and ceased altogether in 1975 and 1976, but were resumed again in a modest way in 1977. Algerian LNG was first imported in very small quantities in 1970 and these have built up rather slowly to the equivalent of 31 million cubic feet per day by 1977, apart from 1974 when no Algerian LNG was imported. Virtually all LNG imports up to 1977 have been for the account of Distrigas of Boston. However, 1978 saw the commencement of LNG imports by El Paso (also from Algeria) and by the end of 1980 or thereabouts, these should have built up to the equivalent of over 1000 million cubic feet of gas per day. Shortly thereafter a third scheme from Algeria—the so-called Trunkline project—is planned to come into operation and this will represent, at plateau rates of delivery, approximately 450 million cubic feet of gas per day.

These three projects—Distrigas, El Paso and Trunkline—are the only LNG import projects which have so far received all the necessary governmental approvals and regulatory consents. Collectively they will amount to approximately 1600 million cubic feet of gas per day at full rates of delivery.

Up to 1957 the United States was a net exporter of natural gas, but in 1958, and in all subsequent years to date, imports have progressively exceeded exports reaching an annual net import level since 1971 of at least 850 000 million cubic feet (over 2300 million cubic feet per day). This trend of increasing volumes of imported gas is expected to continue in the foreseeable future. The Carter administration, or more precisely the DOE, has indicated in various public statements that after maximisation of indigenous gas supplies, the hierarchy the DOE presently (as at early 1979) accords to supplementary supplies is, firstly, Alaskan gas—strictly speaking indigenous gas, but it has to transit Canada—followed by increased imports of gas from Canada and Mexico, both presumably to be delivered mainly or exclusively by pipeline. Next in order of attractiveness is said to be short-haul LNG, although what is classified as short-haul has not been clearly defined; presumably in practice it means proximate prospective supply sources such as Trinidad. Next is synthetic or substitute natural gas (SNG) made from oil and coal feedstocks, depending on the resolution of technical problems and cost factors. Last would be long-haul, high-cost LNG imports, without specifying as yet which prospective supply sources would fall into this category.

In support of the foregoing, James Schlesinger, Secretary of the DOE, stated in January 1979 that the Natural Gas Policy Act of 1978 had: 'removed the pernicious effects of the dual market, which encouraged surplus capacity in the intrastate market while precluding effective competition from the interstate market', and, 'the NGPA replaced cost-based pricing with incentive pricing to encourage additional exploration and drilling through better price incentives and better market opportunities through unconstrained access to the interstate market.'

Evidence of the DOE's attitude to imported gas and their indicated hierarchical approach to such supplies will obviously take time to emerge, but nevertheless the indications are that these policies are in the forefront of their present thinking. For instance, in December 1978 the ERA denied two long-standing LNG import applications—the so-called El Paso II and Tenneco (via Canada) LNG projects from Algeria, both for about 1000 million cubic feet of gas per day over 20 years. It will be interesting to see how other import applications, both for pipeline gas and for proximate and long-haul LNG, are treated by the DOE in the coming years. Much will obviously depend on the degree of success achieved by the industry in finding and exploiting new gas reserves on land and offshore the United States, the relative costs of bringing Alaskan gas to the lower 48 states, and the costs of producing SNG from oil and coal feedstocks.

Natural Gas Liquids

Before leaving the American gas scene, brief mention should be made of the very substantial business that has grown up based on the extraction of natural gas liquids (NGL) from 'wet' natural gas. In 1976 the worth of this business, based on ex-processing plant prices, was assessed to be $3280 million. At end 1976, the number of NGL processing plants in operation totalled 763 with a combined gas processing capacity of 72 610 million cubic feet of gas per day.

Apart from the 'added value' the extraction of NGL offers the producer, it is necessary to remove those constituents of natural gas which would otherwise cause technical problems in the transmission and distribution of natural gas. Having satisfied these technical requirements the extent to which hydrocarbons heavier than methane, in particular ethane, may be further extracted is conditioned by the need to ensure that the calorific value and burning characteristics of the resultant 'dry' natural gas remain within acceptable limits. In 1977, the calorific value of natural gas distributed in the United States after treatment averaged 1021 Btu per cubic foot: generally speaking (there are some exceptions) it cannot be permitted to fall below 1000 Btu per cubic foot.

The liquids (when present) which are extracted comprise ethane, propane, isobutane, normal butane, natural gasoline, isopentane and heavier hydrocarbons (condensate). These liquids are used variously as fuels in their own right (including automotive uses), as refinery fuel, as feedstocks for the petrochemicals industry, as gasoline blending components, etc. Another measure of the significance of the NGL business is that whereas in Western Europe, Japan, and in most other industrial countries, the petro-chemicals industry uses mainly either naphtha and/or gasoil as its feedstock, in the United States natural gas liquids, notably ethane, predominate.

3.8 Gas processing plant for extracting natural gas liquids and sulphur, Waterton, Alberta (*Shell*)

The decline in indigenous natural gas production since 1973 was not mirrored precisely by a corresponding decline in NGL availability because of deeper extraction of gas liquids and improved NGL plant thermodynamic

3.9 A close-up of a gas processing unit at Jumping Pound natural gas liquids plant, Calgary (*Shell*)

efficiency. Even so NGL production has declined since 1973 (see *Table 3.10*), and it will parallel natural gas production much more closely in future as little scope now remains for yet deeper extraction without impairing the quality/calorific value of natural gas.

TABLE 3.10 NGL PRODUCTION IN USA: 1972–76 (MILLION BARRELS)

Product	*1972*	*1973*	*1974*	*1975*	*1976*
Ethane	101	108	118	123	134
LPG (propane and butane)	344	339	330	321	304
Natural gasoline	156	156	144	130	129
Condensates, etc.	37	32	24	22	20
TOTALS	638	635	616	596	587

A large number of NGL export schemes are now in various stages of construction or planning in the Middle East, Africa, the Far East, and elsewhere, and such supplies could theoretically make up any shortfall in local availability of NGL production in the United States. However, whether governmental permission will be forthcoming for such imports in the quantities likely to be required is still far from clear. Governmental policies on NGL imports are as yet even less formulated than for LNG imports.

Some Concluding Thoughts

The United States is the world's largest, most complex and sophisticated gas market. At the same time it is to all intents and purposes totally regulated or circumscribed by Federal and/or state legislation, rather surprising perhaps in a country which has long been regarded as a bastion of free enterprise and enterpreneurialship. All the very many aspects of the gas industry and the legislation pertaining thereto cannot possibly be covered in one chapter of a book of this nature—many of them warrant a book to themselves—and all that has been attempted here is to select some of the more notable events and circumstances that indicate how and why the American gas business has developed in the way it has.

The problem for the author has been to decide what to include and what to omit, and any serious student of the American gas industry would be well advised to extend his reading in order to gain a greater understanding and knowledge before jumping to any premature conclusions based on what has been written here. Nevertheless, in spite of this qualification one fact emerges clearly: it is that the gas business in the United States by its sheer size and importance to the American economy cannot be overlooked, and that it has had a fundamental influence on the development of the country's economic strength and living standards over the past decades. Undoubtedly its influence in future years will continue to be considerable, even though natural gas's contribution to total energy consumption and annual rates of production have declined somewhat since the early 1970s. While this recent trend is unlikely to be reversed, there are many who consider that the decline can be halted, albeit with the aid of increasing imports from Canada, Mexico, and from further afield, supplemented in due course by the manufacture of high calorific value substitute gases from coal and oil feedstocks.

The United States' gas industry is now at yet another cross-roads; the contribution it can make to future energy requirements, the types of market it will be able to supply, the future supply and market mix, and a whole host of other factors that will govern its conduct and future prosperity, remain matters for debate and speculation and as such are not proper subject matter in this historical account.

4: Japan

The gas industry of Japan is now just over one hundred years of age and is at that interesting stage of evolving from a traditional, fragmented, manufactured gas-based industry to a rather more sophisticated business, albeit still fragmented, relying to an increasing extent on imported natural gas in liquefied form. In terms of gas connections, Japan is today one of the world's largest gas markets with nearly 15 million consumers: the market is characterised by a number of unique features as will become evident later in this chapter.

The Early Days

Gas manufactured from coal was first used in Japan for lighting by the Satsuma Clan in the Kagoshima Prefecture of Kyushu in 1857. Fourteen years later, the 4th year of the Meiji era (1871), a gas works was established in the Osaka Mint for melting gold and silver and for minting coins. Surplus gas was used to light some 600 gas lanterns in and around the mint.

In the following year, Japan's first public gasworks was built in Yokohama by Kaemon Takashima under the technical supervision of Pelegrin, an engineer from the Shanghai gas company. This plant comprised a battery of horizontal retorts with a maximum output of about 80 000 cubic feet per day, and a gas holder with a capacity of 30 000 cubic feet. The total length of the distribution mains (8 inch diameter), when the system was completed, amounted to 23 800 metres. Gas prices were yen 3.75 per 1000 cubic feet for indoor lighting, and yen 4.44 a month for outdoor street lights, plus a maintenance service charge of yen 0.115 per month per light. This was the beginning of the Japanese gas industry, some sixty years after the establishment of the first gas company in the United Kingdom.

At a much earlier date, *c.* 1645, natural gas was discovered in the Niigata Prefecture of Honshu. But it was not until 1907 that natural gas was first distributed for public use in Niigata, and not until after the Second World War that locally produced natural gas began to make a worthwhile contribution to Japan's gas supply availability.

Reverting to the development of manufactured gas, in 1874 construction of a gas works in Tokyo was started, again under the supervision of Pelegrin. Initially, the maximum daily production capacity of this plant was 25 000 cubic feet (700 cubic metres). This was expanded to provide a maximum daily capacity of 90 000 cubic feet (2600 cubic metres) by 1877. However, sales developed at a rather slower rate. By the end of 1874 only 85 street lights in Ginza, Tokyo, were being supplied with gas at a price of yen 3.75 per 1000 cubic feet. Although by the end of 1877 sales had expanded to about 4000 cubic feet daily (equivalent to some 40 000 cubic metres per annum) with gas being supplied to 19 houses and 637 street lights, the limited use of gas by the general public caused the company many hardships. This enterprise, which at that time was municipally owned, was transferred to private ownership in 1885, leading to the foundation of Tokyo Gas Company, today Japan's largest gas company.

In 1887, the total volume of gas manufactured and supplied throughout Japan was 650 000 cubic metres, and this rose to 3 620 000 cubic metres in 1897. Although this represented a doubling of gas supply and consumption about every five years over this period, there were still only four gas companies in operation—the municipally owned Yokohama Gas Works, Tokyo Gas, Kobe Gas and Nagasaki Gas. Other gas companies, such as Osaka Gas and Fukuoka Gas, had built manufacturing plant but had yet to start commercial gas distribution operations.

The Post-Russo-Japanese War Period

After the Russo-Japanese War of 1904–5, there was an economic boom which gave rise to flourishing business in almost every industrial field of activity. The number of gasworks which stood at 6 in 1905, increased to 20 in 1909, to 33 in 1911, to 75 in 1912, and to 91 in 1915.

Throughout this period the main use of manufactured gas was for lighting, with only limited use for cooking and industrial applications. The introduction of the electric lamp around the turn of the century was, however, a temporary set-back for the gas industry. In view of similar references in previous chapters, it is perhaps appropriate to digress for a moment on the invention of the electric lamp.

Joseph Swan demonstrated the first electric lamp to a meeting of the Literary and Philosophical Society in Newcastle, England, on 3 February 1879. Thomas Edison also discovered the incandescent electric lamp

4.1 Tokyo's first gas works built in 1874 (*Gas Shiryo-Kan*)

4.2 A gas street light in the Muromachi district of Tokyo, 1874 (*Tokyo Gas Company Limited*)

independently of and at about the same time as Swan. It seems probable that Swan gave the first public demonstration and that Edison secured the first patent. They did in fact collaborate to form the Edison and Swan Electric Light Company in 1882.

These early electric lamps gave a rather poor, soft light, and manufactured gas soon made a come-back as imported incandescent gas mantles began to become available in Japan. These gave a superior light to the pre-tungsten form of electric lamp which followed some years later. A further boost to gas sales was the gas engine, devised by Phillipe Lebon but not introduced on a commercial scale until the 1860s, which offered a number of advantages over the conventional and widely used steam engine. This application for gas was, however, fairly short-lived, and after reaching a peak of nearly 2700 gas-fired engines in 1914, it subsequently declined as small electric-powered motors began to appear on the scene.

Gas-Making Processes

Most Japanese gasworks in the early days utilised the horizontal retort system. Two basic kinds were employed; the so-called stop-end-type horizontal retort system used mainly by the larger gas works, and the through-type horizontal retort system by smaller gas works.

Other methods of gas-making were not popular in Japan where the relatively low cost of horizontal retorts, their ease of construction, and the ability to add quickly further production capacity found favour with the Japanese in spite of their relatively high operating costs and limited coke production compared with some alternative processes.

Some of the larger companies, such as Tokyo Gas and Osaka Gas, devised various technological variations and innovations, and developed processes to produce by-products such as benzene and creosote around the turn of the nineteenth century.

The Gas Industry Goes into Decline

The invention of the tungsten electric lamp in 1910, which then began to find its way into Japan, started the second, and this time irreversible, demise of the use of gas for lighting as it did in almost every other country where manufactured gas was marketed. Gas as a means of lighting and power (for gas engines) began to recede in the face of increasing competition from electricity. Whereas the earlier introduction of incandescent gas mantles had brought the electricity industry near to the point of bankruptcy, the tungsten lamp and the electric motor reversed the trend once again.

So 1914 was a peak year for the gas industry and was not to be reached again for some years to come. In that year the industry comprising 88 companies produced over 171 million cubic metres of manufactured gas and had 643 469 consumers: it was supplying gas to 9525 street lights, and in addition there were 1 544 115 indoor lights lit by gas. In fact gas production increased somewhat the following year to 176 million cubic metres, but the number of consumers dropped by 32 462. By 1920 the total number of gas consumers had fallen to 492 214 and the number of gas companies to 77.

The problems of the gas industry were exacerbated by the more or less coincidental effects of the First World War. Commodity prices, especially the price of coal, shot up, as did wages and transport costs; inflation was rampant. However, because of agreements that had been made with the municipalities, gas prices were virtually pegged, and such increases as were achieved did not cover the full increase in costs and there were delays before they could be implemented.

The gas industry fell into a depression; some companies went bankrupt while others had to enter into mergers. The need for some form of legislation to protect and to help the gas industry to redevelop became critical.

The Post-First World War Era

By the early 1920s the Japanese economy started to recover. Manufactured gas began to be used more widely as a fuel for cooking. The number of consumers and outlets for gas gradually increased. For example, the number of domestic customers rose from 500 000 in 1921 to 790 000 in 1926, of which about 80 per cent were in cities—gas penetration in towns and villages was modest, as it still is today.

The Great Kanto Earthquake Disaster (Kanto Daishinsai) of 1 September 1923, was estimated to have caused the death of about 144 000 people in the Yokohama and Tokyo area, the majority of whom were either burnt or suffocated by the side effects of huge firestones. In addition to this tragic loss of life, the earthquake wrought great havoc to the gas supply systems in these cities and interrupted, for a time, the redevelopment of the gas business which had started a few years earlier.

Legislation

The need for legislation has already been touched upon. The existing so-called bond of remuneration between gas companies and municipalities as a means of regulating gas prices and providing rights of way in urban areas created many conflicts between the two sides. A gas act bill introduced to the 14th Imperial Diet in 1922—a first attempt to change affairs—was shelved.

However, a new and modified bill was passed in April 1923, at the 46th Imperial Diet, with an effective date of October 1925. The main provisions of this Gas Act included:

(1) the establishment of new gas manufacturing plant and modifications to existing plant to be on a permission basis;
(2) regulations governing the use of public property such as rivers, roads, etc., for distributing gas;
(3) the regulation of gas prices and related terms and conditions;
(4) safety aspects for gas manufacturing plant; and,
(5) limitations relating to calorific value, pressure, and quality.

The Act paved the way for creating a healthy gas industry to be conducted essentially as a public utility. Amendments to the Act were necessary in 1930 to reflect the changes that had taken place in Japanese society over the years. These amendments required any increase in the capital structure of a gas company to be on a permission basis, provision for the relevant Minister to rule on unresolved disputes between gas companies and municipalities, and the introduction of standard calorific values on a daily and on an average monthly send-out basis.

The Business Grows

Against the background of a sound legal framework, the Japanese gas industry established a firm foundation and grew steadily. By 1935, the total number of gas consumers or connections amounted to 1 990 000, a four-fold increase since 1921; total sales amounted to 770 million cubic metres. With a then thriving munitions industry the use of gas for industrial purposes expanded rapidly. The outbreak of the Sino-Japanese War in July 1937 increased yet further the demand for gas. In 1938, sales reached 970 million cubic metres and the total number of consumers stood at 2.3 million, supplied by 118 gas companies.

As the war became protracted the government found it necessary to introduce measures regulating the production and consumption of various commodities to ensure the war effort was not impeded. For gas this meant restrictions on the procurement of coal and, in 1939, regulation of the amount of gas consumed by domestic customers.

Second World War

Japan entered the Second World War on 7 December 1941. The step-up in war effort led to further restrictions. The shortage of feedstock (coal) for the gas industry resulted in prohibition of the use of gas for domestic purposes other than cooking, and taxes on gas (10 per cent) and electricity were imposed in order to suppress consumption and to raise finances for war expenditure. Allocations of the monthly permitted consumption of gas per household were laid down which varied with the size of the family unit. Authorisations were given to gas companies in 1942 to discontinue supplies for a set period if a family consumed more gas than its monthly entitlement.

Although the domestic gas consumer was hard hit, nevertheless he was perhaps rather more fortunate than others not connected to the gas grid who had to forage for scarce supplies of wood, charcoal and coal. Towards the end of 1942, gas supply was yet further restricted with a 15 per cent cut for domestic uses, 30 per cent for commercial and industrial uses of a non-essential nature, while for offices supplies were discontinued other than for essential purposes.

These restrictions, followed in due time by the consequential effects of air raids, distort unduly statistical comparison of the pre-war period with war-time days. Variously 60 to 90 per cent of the homes of domestic gas consumers in such cities as Tokyo, Yokohama, Osaka and Hiroshima, were destroyed. Such statistics as are available, which undoubtedly are rather imprecise, indicate that the number of gas consumers throughout Japan during the war dropped to perhaps 1.4 million compared with over 2.4 million in 1941.

Gas consumption data are even more meaningless as, because of the damage done to the distribution system, gas leakages exceeded 50 per cent in some areas and were probably as high as 30 per cent for Japan as a whole. The Japanese gas industry in one way or another suffered badly as did, of course, the gas industries in some other countries over which war was raged; nevertheless it recovered remarkably quickly.

The Immediate Post-War Years

Reconstruction of the industry started shortly after the war. By November 1945, gas was being supplied again even to the most devastated areas. Leakage on a nation-wide basis was fairly quickly cut from 30 to 15 per cent, a not inconsiderable achievement. Gas manufacturing plants, transmission systems, distribution mains and all related facilities, were repaired or renewed as fast as possible.

In contradistinction, the post-war coal shortage worsened and all gas companies were obliged to restrict gas supplies. Some companies were in consequence obliged to supply gas only for a few hours in the morning, at lunch time, and in the evening. As coal rationing became tighter, supplies were further reduced to two or three hours a day and/or suspended on Sundays: legal restrictions were enforced on the use of gas in 1946.

Things gradually improved and in 1949 governmental regulations controlling the use of coal were lifted, except for imported coal and for certain types of coal and coke required for special purposes. At the same time gas was decontrolled and the industry was once again able to resume a basic 24-hour supply to its customers. In essence, the stage was set for the gas industry to resume its proper role after some years of restriction, curtailment of business, and immense practical problems essentially outside its own control and making.

New Legislation

In 1950, a Public Utilities Ordinance was enacted and the Gas Act was repealed. This embraced both the gas and electricity industries, and a Public Utilities Committee was organised to co-ordinate the activities of both industries. This committee was formed on a collegiate system, being independent from the administration but responsible to the legislature. For the first time in Japan public hearings were held and the collection of opinions undertaken. However, following the invalidation of the Ordinance in 1952, the committee was dissolved.

This was followed in some haste two months later by a law containing provisional arrangements governing the gas and electricity industries. It seems that the Japanese, at that time, were trying to mirror the role of the United States Federal Power Commission, without perhaps recognising at first the disparities of these industries in the two countries concerned. This led to another act which was passed by the Diet on 31 March 1954 and which, inter alia, reinforced regulations protecting consumers' interests, emphasised the gas industry's obligations, required certain matters to be subject to permission or licence, amended various necessary minimum requirements, and yet at the same time respected the independence of the gas industry.

At last legislation had been established, appropriate to the times and the Japanese style of business, which contained the right blend of checks and balances so as to ensure that the gas industry developed in an orderly and responsible manner, but without stifling its initiative and growth potential.

New Feedstocks and Natural Gas

In 1952, Tokyo Gas Company produced gas from oil feedstocks and became the leader of the Japanese gas industry in the conversion from coal to oil gas-making processes. Statistics for the following year show that the average annual consumption per connection was 2520 megacalories (Mcal) compared with 1560 Mcal in 1941—here it is necessary for the third time in this account to switch to another unit of measurement in line with the changes in published statistical data—and that gas was being used to an increasing extent for bathroom water heaters and for room space heating as living standards improved.

A radical change that took place in the late 1950s/early 1960s was the construction of new gas-making plants using locally produced natural gas, crude oil, naphtha, and LPG as feedstocks, and the gradual conversion of the existing coal-based manufacturing plants to these newer types of feedstock.

Available reserves of indigenous natural gas were, and still are, small in relation to total demand. Although production has increased every year since the mid-1950s, by 1968, the year before the commencement of LNG imports, locally produced natural gas contributed only 3800 teracalories (Tcal), or 9 per cent, to a total supply of all types of gases used by the gas industry amounting to 40 900 Tcal. In fact in that year (1968) total production of natural gas amounted to some 23 850 Tcal, but over 60 per cent of this was supplied directly to the chemical industry, a further 16 per cent was absorbed for power generation and by other industries. By 1976, indigenously produced natural gas supplied to the gas industry had risen to an annual rate of 6500 Tcal, but with the overall growth in gas supply and consumption natural gas's share of total supply had by then dropped marginally to 8 per cent. In the absence to date—in spite of a considerable exploration effort on land and offshore—of any major discoveries, it was impossible for Japan to consider indigenous natural gas as anything more than a useful supplement to manufactured gas, and more recently, of course, as a supplement to imported liquefield natural gas as well.

Let us, therefore, now examine some of the changes that have taken place over the most recent 10-year period for which detailed statistics are available—from 1968 to 1977.

Customer Growth and Consumption Trends

As far as the number of customers is concerned, no other gas market in the world has grown at such a phenomenal rate in both absolute and percentage terms over the period 1968–77. This is illustrated in *Table 4.1*, which shows that the number of customers, principally domestic consumers, increased by 6.2 million, equivalent to an average annual increase (a.a.i.) of 6.1 per cent; the comparable figures for the very much larger United States' market were over 5.3 million customers and 1.3 per cent a.a.i.

TABLE 4.1 NUMBER OF GAS CONNECTIONS/ CUSTOMERS: 1968 AND 1977 (000s)

Market sector	*1968*	*1977*	*a.a.i.* %
Domestic	8 222	13 998	6.1
Commercial	372	719	7.6
Industrial	45	60	3.2
Others	80	131	5.6
TOTAL	8 719	14 908	6.1

The quantities of gas sold by town gas undertakings, including both manufactured and natural gases, but excluding direct purchases of gas (mainly imported LNG) by electricity generating companies, have also increased dramatically as shown in *Table 4.2*.

TABLE 4.2 SALES BY GAS COMPANIES: 1968 AND 1977

By main market sector	*1968* (*Tcal*)	*1977* (*Tcal*)	*a.a.i* (%)
Domestic	22 056	49 632	9.4
Commercial	7 031	13 959	7.9
Industrial	4 292	9 648	9.4
Others	2 120	4 554	8.9
TOTAL	35 499	77 793	9.1

From *Tables 4.1* and *4.2* the average rates of consumption shown in *Table 4.3* can be derived:

TABLE 4.3 AVERAGE ANNUAL CONSUMPTION: 1968 AND 1977

By type of consumer	*1968* (*Mcal*)	*1977* (*Mcal*)	*a.a.i* (%)
Per domestic consumer	2 682	3 546	3.0
Per commercial consumer	18 901	19 414	0.3
Per industrial consumer	95 378	160 800	6.0
Others	26 500	34 763	3.0

Although the average annual consumption per domestic customer has risen over the last 10 years by about 1 gigacalorie (1000 Mcal) this is still remarkably low compared with many other countries. For example, the average consumption per domestic customer in the UK in 1977 was 11 700 Mcal, and in the United States 29 200 Mcal. This wide disparity is due primarily to the fact that until relatively recently most of the houses in Japan were built of wood and heated mainly by kerosene, and to a lesser extent by LPG, solid fuels and electricity. The principal use of town gas has been, and still is for that matter, cooking, with water heating being the next most important application. However, apartment blocks and similar multi-family units are becoming an increasing proportion of the new housing now being built in the larger Japanese towns and cities. These are built of concrete and similar materials, for which the sealed, balanced-flue and forced-flue, gas-fired central heating systems have obvious technical and environmental advantages over the older heating systems that traditionally have been used in wooden houses for many years past.

4.3 Laying a new distribution main in Osaka, Japan, 1978 (*Osaka Gas Company Limited*)

In spite of these advantages, and the increasing availability of natural gas, the pattern of fuel usage in the domestic market is only changing very gradually. This is because the present tariff (pricing) structure is not economically attractive for domestic consumers. Until such time as the tariffs are changed, penetration of the space-heating market will remain a slow progress, as also will be the case in developing the potential market for gas-fired air-conditioning, where there is the added problem of the relatively high

4.4 The distribution control centre of Osaka Gas, Japan, 1978 (*Osaka Gas Company Limited*)

capital cost of equipment to be overcome. However, even if the domestic tariff structure is changed so as to facilitate and encourage the use of gas for space heating and air-conditioning, it is unlikely that average consumption rates would ever reach those achieved in the United States or in northern Europe, if only because of the less severe climatic conditions that generally prevail in Japan.

As will be seen from *Table 4.2* the commercial sector (i.e. supermarkets, schools, hospitals, hotels, restaurants, and other non-industrial large consumers of gas) is an important outlet for gas in Japan. The demand for gas in this sector for space heating and air-conditioning has increased in spite of the relatively unfavourable tariff structure. This is largely caused by a combination of legislative restrictions which favour the use of non-polluting fuels, and/or the fact that storage space at customers' premises for alternative liquid fuels is frequently not available in the more congested inner-city areas. While there has been an encouraging upward trend in the numbers of new commercial customers and in their average annual rates of consumption in recent years, it is doubtful if this trend can be maintained unless changes to the tariff structure are introduced so that gas can compete on price with other fuels in the less congested areas.

Comparison of the industrial market in Japan with markets elsewhere is distorted by the fact that certain industrial customers, in particular some power plants and steel-making companies, purchase imported LNG on a c.i.f. or ex-ship basis for their own account and are not supplied by local gas distribution companies as usually happens in other countries.

The statistics given in *Table 4.2* for industrial consumption exclude these major outlets and thus are not representative of the total use of gas by industry in Japan; this was considerably higher than the 9648 Tcal recorded for 1977. The situation is further confused by the fact that a number of these companies also import LPG directly, which tends to blur the assessment of the total amount of gaseous fuels consumed for industrial purposes. To illustrate the point, Tokyo Electric imported directly for their own account over 65 000 Tcal of LNG from Alaska, Brunei and Abu Dhabi in 1977, more than six times the amount of gas sold directly by gas companies to industrial users.

LNG for Power Generation

This leads us on to another interesting feature of the Japanese gas market, namely that the use of gas, in particular imported LNG, for power generation is a significant and growing market. This is quite different from the situation in Europe and the United States, where the current trend is to phase out natural gas for power generation because, inter alia, this is regarded as an ignoble use of a 'premium' fuel by many governmental authorities and international agencies. In these countries power generation is also one of the lowest value outlets for gas. All this illustrates the point that the markets and preferred uses for gas can and do vary from one country to another.

There are several reasons for this apparent contradiction between Japan and most other countries. The main ones are that in Japan the air pollution regulations applicable in and around the main centres of population are among the most stringent in the world. Electric power companies wishing to increase their generating capacity are in consequence obliged to use non-polluting fuels. As the nuclear option is hardly feasible for power plants located in or close to densely populated areas, this really leaves the choice to nil or very low sulphur content fuels such as LNG, naphtha, and LPG, all of which have to be imported in the absence or insufficiency of indigenous supplies of crude oil and natural gas. An advantage of LNG, over the

4.5 *Khannur*, capacity 125 000 cubic metres, owned by Gotaas-Larsen Inc. This LNG carrier was built to the Moss-Rosenberg spherical tank design, was completed in 1977, and is employed on the Abu Dhabi–Japan LNG project (*Gotaas-Larsen Inc.*)

4.6 Sodegaura LNG reception terminal, Tokyo Bay, Japan (*Tokyo Gas Company Limited*)

relatively few alternatives available, is that LNG supplies can be secured under long-term (15 to 25 years) contracts on a negotiated pricing basis. Provided LNG is competitively priced with such relatively few alternative fuels as can be considered, LNG is probably a more reliable source of supply in that the alternatives tend to be only available against shorter duration contracts and are subject to more frequent price variations as determined by the vagaries of the international oil business. LNG, which is a non-polluting fuel, is almost invariably supplied on a regular long-term basis; it can be stored conveniently and used when required and is an eminently suitable fuel for power generation in the somewhat special circumstances pertaining in Japan. For these and other reasons, natural gas in the form of LNG is likely to be used to an increasing extent for power generation and this market is expected to remain in the foreseeable future the principal outlet for gas in Japan.

The pattern of direct imports of LNG by power companies has already been set, and a similar pattern may develop with certain other prospective major users, in particular the steel industry, in the years ahead. To the extent that this trade in direct purchases of imported LNG develops, it will distort the pattern of development of the industrial market by Japanese gas companies. Sales by the latter to the so-called industrial market will tend to continue to be confined to selling gas to relatively small and medium-sized industrial consumers, mainly for high grade applications, who, because of the volume of their requirements or because of their geographical location, cannot contemplate purchasing imported supplies direct for their own account. In these circumstances the main volumetric outlets for Japanese gas companies are, and are likely to remain, the domestic market supplemented by the commercial market and, where appropriate, by high-grade industrial applications—inevitably there will be some important exceptions to this general trend.

Types of Gas Distributed

In terms of combustion characteristics there are something like 14 different types of gas distributed today in Japan, apart that is from the direct sale of LPG tel quel (butane, propane and/or mixtures thereof) to domestic, commercial, industrial and other users such as the chemical industry. The calorific value of the gas supplied ranges from about 3600 kilocalories per cubic metre (kcal/m^3) to 11 000 kcal/m^3, or to 24 000 kcal/m^3 if one includes the piped distribution of LPG undertaken by four gas companies. This is probably the widest range of calorific values for any gas market in the world and points up the fact that there are a large number (252 in 1977) of independently operated companies, also the absence, as yet, of any regional or national supply system or pipeline grid.

The types of gas distributed during the last 10 years can be classified under the following main groupings: coal gas; producer gas; water gas; gas made from oil feedstocks (crude oil, kerosene, naphtha, and including refinery tail gases); LPG (straight or mixed with air); locally produced natural gas; and imported LNG. Some of these main categories could, of course, be further subdivided, but this may confuse rather than aid understanding. From a peak of 7.1 million tonnes in 1970, the use of coal as a gas-making feedstock has slowly declined to 5.1 million tonnes in 1977. Likewise over the period 1968–77 coke and crude oil have also declined progressively each year (e.g. for coke from 97 000 tonnes in 1968 to 24 000 tonnes in 1977). Kerosene, naphtha, and LPG have all increased (e.g. LPG from 125 000 tonnes to 722 000 tonnes), but the most dramatic increase has been LNG—from 19 000 tonnes in 1969, the first year of imports, to 2.1 million tonnes in 1977. All these figures are, of course, in terms of the amounts of feedstock used for gas-making. *Table 4.4* shows the increases or decreases in the resultant gases available for distribution.

Number of Gas Undertakings

The wide variety of types and calorific values of the gases distributed reflect the fragmented nature of the Japanese gas industry. In 1977 there were 179

TABLE 4.4 GASES AVAILABLE FOR DISTRIBUTION: 1968 AND 1977

By main classifications	*1968* (*Tcal*)	*1977* (*Tcal*)	*a.a.i./d.* (%)
Coal gas	16 393	15 064	(0.9)
Producer gas	457	124	(13.5)
Water gas	85	nil	—
Crude oil gasification	9 766	6 387	(4.8)
Kerosene/naphtha gasification	8 740	19 608	9.4
LPG	1 528	8 730	21.5
Natural gas (indigenous)	3 833	6 713	6.8
Regasified LNG (imported)	nil	27 394	—
Others	61	82	2.3
TOTAL	40 863	84 102	8.4

Note: The differences between these totals and the sales figures given in *Table 4.2* represent mainly own use and losses.

privately owned and 73 municipally owned gas undertakings, of which 130 had less than 5000 consumers each. Indeed there were 19 undertakings which individually had less than 1000 consumers. At the other end of the scale there were only four companies (all privately owned) with over 500 000 consumers per undertaking. These were the Tokyo Gas Company, the world's largest privately owned gas company with about 5.5 million consumers, followed by Osaka Gas Company (nearly 4 million), Toho Gas Company (over 950 000) and Saibu Gas Company (around 650 000). In 1977, about 73 per cent of the total gas sold by distribution companies and about 70 per cent of the aggregate number of gas customers in Japan were supplied by the three leading gas companies of Tokyo, Osaka and Toho.

Conversion

Conversion from manufactured gas to natural gas, based primarily on imported LNG, is going ahead in Tokyo and Osaka, and got underway in Nagoya during 1978. In Tokyo, conversion to natural gas started in 1972 and by end 1978 roughly 50 per cent of the existing consumers had been converted. Current planning envisages completion of the conversion programme in Tokyo in about 1985. Conversion in Osaka started in 1975 and is likely to take 10 years to complete. While this may appear to be an inordinately long period compared with North American and European experience, the circumstances are not directly comparable. Cities such as Tokyo and Osaka are more congested than their counterparts elsewhere and conversion has to proceed at a somewhat slower pace if major disruptions are to be avoided. In addition, the programmes for imported LNG take time to establish and for supplies to build up. They also involve the creation of a far more complex infrastructure than is the case with pipeline supplies which most other countries have had as a basis for their decision to convert to natural gas.

In order to maximise the energy-carrying capacity of their existing distribution system, Tokyo Gas and Osaka Gas have decided to enhance the calorific value of their regasified LNG, which is currently about 10 000 kcal/m^3, by the addition of LPG so as to be able to distribute a resultant gas blend with a gross calorific value of 11 000 kcal/m^3. Apart from increasing the capacity of their existing system, this enhancement of calorific value should have the added merit of giving these companies greater flexibility in being able to accept a somewhat greater range of gases of differing calorific values from a variety of sources in the years ahead.

The Introduction of LNG

Frequent references have already been made to the growing importance of LNG supplies for the Japanese gas market. In 1966, Tokyo Gas and Tokyo Electric concluded 15 year contracts with two American companies, Phillips Petroleum and Marathon Oil, for the importation of nearly 1 million tonnes per annum of LNG to be supplied from a liquefaction plant to be constructed at Port Nikishi, Cook Inlet, Alaska. The first delivery against these contracts arrived at Tokyo Gas's reception terminal at Negishi, Yokohama, on 4 November 1969. This project, to which two LNG carriers each of 71 650 cubic metres (m^3) cargo capacity are dedicated, has operated successfully since its inception. Some 75 per cent of the annual delivered quantities are utilised for power generation by Tokyo Electric with the balance going to Tokyo Gas for public distribution purposes.

In June 1970 contracts were signed between Coldgas Trading Ltd, a joint Shell/Mitsubishi company, and Tokyo Gas, Tokyo Electric, and Osaka Gas, for the supply over 20 years of some 3.7 million tonnes per annum of LNG from Brunei. The first delivery against these contracts was made to Osaka Gas's terminal at Senboku on 15 December 1972. Before these supplies commenced the annual contractual quantities were raised by 1.5 million tonnes to a total of 5.2 million tonnes per annum. Seven LNG carriers, with cargo capacities in the range of 75 000 to 77 000 m^3, are dedicated to this project and deliver LNG to three reception/regasification terminals at Senboku, Negishi and Sodegaura, the latter two being located in Tokyo Bay. In 1977, certain corporate changes were made with Brunei Coldgas, a company owned as to one-third each by Shell, Mitsubishi and the Brunei government, taking over the contractual responsibilities previously vested in

Coldgas Trading. This project has also operated successfully since its inception in 1972.

In chronological order, the next LNG import project was from Das Island, Abu Dhabi. This venture is for the supply of 2.2 million tonnes of LNG per annum to Tokyo Electric over 20 years with the first delivery being made to Sodegaura on 14 May 1977. On the supply side the companies involved are the Abu Dhabi National Oil Company, BP, CFP, Mitsui and Bridgestone. The project employs four LNG carriers with capacities ranging from 87 500 to 126 400 m^3.

A more recent project to come on stream is based on Badak, East Kalimantan, Indonesia, which delivered its first cargo to Osaka on 14 August 1977. This project, together with a second Indonesian scheme based on Arun, North Sumatra, which delivered its first cargo to Chubu Electric at Chita on 12 October 1978, will involve a total supply at plateau rates of delivery of 7.5 million tonnes of LNG per annum over 20 years to a group of Japanese customers comprising, Kansai Electric, Chubu Electric, Toho Gas, Kyushu Electric, Osaka Gas and Nippon Steel. Pertamina, the Indonesian state entity, has a major interest in both projects, in partnership with the Huffington group in the case of Badak and Mobil Oil in the case of Arun. Seven 126 750 m^3 capacity LNG carriers will in due course be employed in transporting these supplies from Indonesia to Japan.

It is to be hoped that when these more recent import projects reach their full operating capacity they will perform to the same degree of reliability that has been the case with the Alaskan and Brunei projects.

The Role and Future Scope for LNG

As already stated the decision has been taken, at least by the three major gas companies, to convert not only to natural gas, but more specifically to imported liquefied natural gas. As Japan has long relied on importing the majority of the fuels and raw materials it requires perhaps this is not a surprising decision, particularly in view of the good experience it has enjoyed to date with LNG imports from Alaska and Brunei. And the fact that Japanese gas companies and other major users of gas are actively seeking additional supplies of LNG from a variety of sources is evidence of the confidence they hold in LNG. By the mid-1980s, existing supplies should be supplemented by further supplies from Indonesia and with LNG from Sarawak and Australia ; other potential LNG supply sources in the Middle East, South-East Asia and elsewhere could well materialise in the latter part of the 1980s or the early 1990s.

Thus although the Japanese gas industry is planning to commit itself whole-heartedly to LNG, at the same time it will have access to an increasing diversity of supply sources with no undue dependence on any particular project, country, or geographic region. Strategically this makes good sense. In all cases only a part of these supplies will actually go into the gas industry's supply system as the major part is expected to be imported and utilised direct by the power generation industry. However, most LNG schemes will probably be undertaken on a co-operative basis with gas companies and other importers sharing common reception facilities. In this way the size and frequency of deliveries (cargo lots) can be optimised to the benefit of all concerned. This should also help to minimise overall storage requirements and to reduce the number of separately located reception terminals in a country where suitable sites are few and far between.

Because of the inevitable uncertainties as to the rate of growth in LNG, and in order to diversify still further its supply sources, the Japanese gas industry is actively developing supplementary sources of high calorific gas, in particular SNG (synthetic natural gas) manufactured from oil (or coal) feedstocks. Various processes are being investigated and several pilot plants are already in operation. When full-scale commercial plants are available, SNG will also help the industry to meet its obligations during peak periods of demand which in a typical winter month can be twice the minimum monthly demand in summer. If the space-heating business builds up, this ratio between summer and winter demand is likely to widen still further.

Summary

The Japanese gas industry is unique in several respects. It contains within it the world's largest privately owned gas company and probably, although this is not possible to confirm categorically, the world's smallest as well. It utilises a wide range of feedstocks and distributes an even greater variety of types and qualities of gas. While the industry is increasingly moving away from manufactured to natural gas in the form of imported LNG, conversion is taking place at a somewhat slower rate than has been achieved in most other countries. The largest individual sales outlet is the domestic cooking and water-heating load, although this may change as other domestic uses, notably space heating, begin to find favour. The gas industry remains fragmented and in general, with a few exceptions, does not supply large, base-load industrial consumers which remain responsible for their own supplies—again in the case of gas, essentially imported LNG. These are but some of the special characteristics of the Japanese gas scene.

The gas industry in Japan in all its many facets is being developed and expanded with customary Japanese vigour, enthusiasm and efficiency. We shall see many changes, not just in relation to gas quality and supply sources, but in a whole host of other matters in the coming decade and it will be instructive at that time to review the progress that has been achieved.

5: The Netherlands

The history of the gas industry in Europe is long and distinguished. The story of gas in the United Kingdom has already been described, and while space precludes similar accounts of how the business evolved in Austria, Belgium, France, Germany, Italy, Spain and elsewhere in Europe, one country that cannot be glossed over is the Netherlands. Although the Netherlands is a fairly small country in terms of its geographic size and population, the contribution that Dutch gas has made to the development of the gas industry in north-west Europe over the last 10 to 15 years has been significant enough to merit selection of the Netherlands for individual treatment in this book.

The event that resulted in the Netherlands becoming such a vital factor in the affairs of the European gas industry in recent years was the discovery in 1959 of the Groningen gas field. Although Groningen is not the world's largest gas field, it is probably the largest accumulation of natural gas that has so far been found in such a favourable location. Within a radius of 250 to 300 miles from Groningen lies much of the industry and population of West Germany and France and the whole of that of the Netherlands and Belgium; indeed northern Italy is barely 600 miles from Groningen. For these reasons the main theme of this chapter is focused on the discovery and subsequent development of Groningen gas, not only for the local Dutch market, but for export as well. First, however, a brief account of earlier days is necessary in order to set the scene for the main story.

Early Developments

One of the first recorded uses of manufactured gas in the Netherlands was for lighting the inside of Ridderzaal, or Knights' Hall, in The Hague in 1820—the building which now houses the Dutch Parliament. However, it was not until

5.1 Rotterdam gas works *c.* 1885 (*N.V. Nederlandse Gasunie*)

1825 that the first privately owned gas company was established in Amsterdam: the second company to be founded was in Rotterdam in the following year. These companies were financed mainly by British interests and capital. This was hardly surprising given that the gas industry in England was already well established and was pioneering the spread of the business in various other countries around the world.

Initially, and in common with the gas industry elsewhere, the main use of manufactured gas in the Netherlands was for lighting public buildings and streets. As gas was expensive in these early days, its ability to penetrate the domestic market was confined to the higher-income members of the population who used it for home-lighting purposes on a limited scale.

Moving on in time, by 1876 manufactured gas enterprises existed in virtually all large Dutch cities and towns and in a number of smaller population centres as well. In that year it is recorded that Bolsward, Friesland, had the smallest gas works with 77 customers and an average annual consumption of 266 cubic metres per connection, and Utrecht the largest with 3700 customers and an average consumption of 755 cubic metres per connection. In The Hague, which had a population in 1876 of about 100 000, there were 2912 gas consumers and 1495 street lights lit by gas.

Improvements both in gas-making processes and appliances, and the invention described elsewhere in this book of the incandescent gas mantle, led to the increased use of gas for lighting and cooking. By 1920 the number of gas companies in the Netherlands had grown to 195. However, soon after this the number of gas companies began to decline, especially in the southern provinces of Limburg and Noord Brabant where traditional gas works suffered severe competition from coke-oven gas produced by the Dutch State Mines. This also happened as a consequence of the increasing availability of blast-furnace gas from the complex of blast furnaces and steel works of Koninklijke Hoogovens en Staalfabrieken at Velsen near IJmuiden; this gas spread ultimately over a large part of the province of Noord Holland, including the main centres of Amsterdam, Haarlem and Alkmaar.

As the years went by, gas lighting increasingly lost ground to electric lighting in much the same sort of way as happened in other countries. By 1938 the pattern of consumption had changed to the point where only 2 per cent of gas supplies were used for lighting: the main market was domestic water heating (71 per cent) and the balance comprised cooking (4 per cent) and industrial applications (19 per cent).

The Immediate Post-War Period

The Second World War hit the gas industry very hard and by the end of it only 150 gas works remained—gas lighting had almost died out. A lot of gas works

5.2 Gas street lights at a cross-roads in Kralingen, a suburb of Rotterdam, *c.* 1900 (*N.V. Nederlandse Gasunie*)

5.3 Preparation of cast-iron pipe before laying. These pipes were connected by spigot and socket joints and made gas-tight with lead and oakum (*N.V. Nederlandse Gasunie*)

were badly damaged or poorly maintained during the war and many of these had to be rebuilt.

An interesting feature of the post-war period was the growth in the use of refinery gases as a raw material or feedstock for the manufacture of town gas throughout the province of Zuid (South) Holland based on the expanding complex of oil refineries in the Rotterdam area. Expansion of the pre-war coke-oven gas distribution systems, based on supplies from Hoogovens and

5.4 The four-stroke Otto gas engine, *c*. late nineteenth century. This engine had an efficiency of about 25 per cent (*N.V. Nederlandse Gasunie*)

the Dutch State Mines, supplemented by some imports of coke-oven gas from West Germany, also took place during these years.

To sum up, by the early 1960s, just before the introduction of Groningen natural gas, the following broad pattern of gas supply and distribution had been built up in the Netherlands. The requirements of the provinces of Noord Holland, Limburg, Noord Brabant and Zeeland were met largely with low calorific value coke-oven gas, as also was Gelderland with imports from Germany. In contradistinction, the provinces of Groningen, Friesland, Drenthe, Overijssel, Utrecht, and in part Gelderland and South Holland as well, had from the early 1950s switched over progressively from manufactured gas to natural gas. In some locations this was distributed tel quel, while in others natural gas was reformed to a lower calorific value quality gas: we shall revert in more detail to the discovery and development of various natural gas deposits in the pre-Groningen period later in this chapter. Finally, the province of South Holland had, as already mentioned, access to large quantities of refinery gases in addition to some locally produced natural gas. And, of course, there were also a number of relatively small gas works scattered throughout the Netherlands making town gas from coal and oil feedstocks supplying localised areas.

In order to put the above into perspective, in 1962, the year before the start of Groningen gas supplies, total distribution of all types of gas (excluding bottled LPG) for public consumption in the Netherlands amounted to a little over 1000 million cubic metres of natural gas equivalent. About 25 per cent of this was natural gas of which about one-half was either reformed to a lower calorific value gas before distribution or used for the enrichment of poor quality manufactured gas. Refinery gases, which were also reformed or used for enrichment, constituted about 20 per cent of total gas supply. Most of the balance was coke-oven gas; coal gas by then having largely but not entirely disappeared. The total number of gas customers at that time amounted to nearly 2.5 million—mostly domestic consumers using gas for cooking and water heating—of which less than 400 000 had been converted to natural gas. Sales to industrial consumers were a small proportion of total gas consumption.

Having set the scene very broadly, let us now turn our attention to the development of Groningen gas for which we have first to retrace our steps to the early part of the nineteenth century when an important piece of legislation was enacted.

The Mining Act of 1810

The general framework of the search for hydrocarbons in the Netherlands was provided by the Netherlands Mining Act, enacted on 8 November 1810, entitled: 'Loi concernant les Mines, les Minières et les Carrières du 21 Avril

1810', founded on French legislation inspired by Napoleon. This Act in essence allowed exploration without licence subject only to the consent of the surface owner. A Deed of Concession by the Crown to the surface owner, to the finder, or even a third party, was, however, necessary before any discovery could be mined.

The Mining Act of 1810 also stated that if the right to mine was not granted by the Crown to the finder the latter was entitled to certain payments to reimburse him for the costs he had incurred, plus a small reward. The surface owner had to be satisfied with a lump-sum payment of guilders 12.5 per hectare, or an annual payment of half a guilder per hectare. If and when actual mining took place the surface owner then became entitled to payment for the use of his land.

The above still forms the basis of Dutch mining legislation, although it has, of course, been considerably expanded and altered in many respects since 1810.

The Foundation of NAM

The N.V. Koninklijke Nederlandsche Maatschappij tot Exploitatie van Petroleumbronnen in Nederlandsch-Indie, later to become known simply as 'Royal Dutch', was founded in 1890. The joining of Royal Dutch and the English firm, the Shell Transport and Trading Company Limited, in 1903, led to the creation of the Royal Dutch/Shell Group of Companies which subsequently set up N.V. De Bataafsche Petroleum Maatschappij (now Shell Petroleum N.V.) better known as BPM.

In September 1947, BPM and the Standard Oil Company of New Jersey (Esso) jointly established N.V. Nederlandse Aardolie Maatschappij (NAM) to explore for and to develop any hydrocarbons it might discover in the Netherlands. These companies, especially NAM, were to play a significant role in the discovery and development of natural gas in the Netherlands as we shall see.

The Search for Hydrocarbons

Between 1903 and 1933 the only prospecting carried out in the Netherlands was by the State Mineral Exploration Service. Apart from finding a trace of oil near Winterswijk, Gelderland, the search was unsuccessful.

In 1933, the search was taken over by BPM, based on a temporary amendment to the law allowing exclusive exploration licences to be issued for certain specified Dutch provinces for a period of 10 years, extended during the Second World War to 20 years. However, in 1947, this system was abandoned in favour of a return to the traditional principle of licence-free prospecting.

The first discovery of oil in commercial quantities was made by BPM near Coevorden, Drenthe, in 1943. This was followed by the finding of more significant deposits near Schoonebeek. But it was not until June 1948 that the first discoveries of natural gas were made, also near Coevorden at a depth of 9100 feet.

In July 1945, BPM submitted an application for a production concession for Schoonebeek under the 1810 law. This was issued in 1948, not to BPM, but at the latter's request to NAM. This concession set the tone for all subsequent concessions granted to NAM and remained virtually unchanged until the organisational upheaval brought about by the discovery of Groningen gas occurred. Successful discoveries of natural gas over this period included De Wijk (1949), Staphorst (1950), Wanneperveen and Tubbergen (1951), Denekamp (1952), Rossum (1955), De Lutte (1956), and Schoonebeek (1958)—all these being in the eastern part of the country fairly close to the German border.

Direct government involvement in the exploitation/production phase was not pressed, but NAM was obliged to pay to the state a proportion of the profits it generated. Also following the 1948 gas discoveries new legislation relating to natural gas was introduced under which NAM undertook to sell to the state-owned entity, Staats Gasbedrijf (State Gas Board), over a 20-year period all gas surplus to own-use requirements. A formal agreement between NAM and the State Gas Board to this effect was signed on 9 July 1948. This arrangement also applied to subsequent gas discoveries in the eastern part of the country and in the province of South Holland.

The Introduction of Natural Gas

Coevorden, a small town at that time of about 5000 inhabitants lying close to the German border, was the first town to receive natural gas for public supply in the Netherlands: this was on 4 September 1951. By the end of 1951, 1300 connections had been converted from manufactured gas to natural gas. In 1952, the State Gas Board started construction of gas transmission grids in the northern provinces. During the remainder of the 1950s many areas, first in the east of the country, and after in the centre and west, received natural gas. In some places straight conversion to natural gas was undertaken; in others natural gas was reformed to a lower calorific value quality gas. The largest town to change to straight natural gas was Leeuwarden (85 000 inhabitants) in 1957, and the largest one to use reformed natural gas was Utrecht in 1959—260 000 inhabitants, but in fact involving a supply area of some 500 000 people.

Clearly natural gas, although its ultimate potential was not suspected at this time, was starting to become an important economic factor in the Netherlands. By 1953, a 'Natural Gas Advisory Committee' recommended that the guideline for the marketing of natural gas should be: 'the highest net profit to the national economy'.

In 1954, the State Gas Board was accorded a monopoly for the wholesale marketing of natural gas throughout the Netherlands. This entity had the obligation to pay for whatever quantities of gas NAM offered it: gas paid for but not immediately taken was held in reserve for subsequent offtake.

The various arrangements described above governing the disposal of Dutch natural gas remained in force until it became clear that the reserves of the Groningen gas field were so large that it would be necessary to make changes in the then existing marketing mechanism.

The Discovery of Groningen Gas Field

The first indication of the existence of the Groningen gas field came on 22 July 1959, when as a result of exploration drilling near the village of Slochteren in the province of Groningen—hence the reason why the field is sometimes called the Slochteren gas field—NAM announced the discovery of a gas-bearing formation. NAM's second exploration well, drilled nearby, confirmed that a substantial deposit of gas had indeed been found, but it was not clear at that time whether this was a fairly localised accumulation or part of a gas-bearing formation extending over a larger area.

Initially, reserves were assessed to be fairly modest at about 60 milliard cubic metres, but were increased progressively as more drilling was carried out and further geological data were gathered. On 11 July 1962, the Dutch Minister of Economic Affairs indicated publicly that reserves were at least 150 milliard cubic metres and, he said, it was possible that they might amount to 400 milliard cubic metres. And on 5 October of that year he updated his previous reserves figure to 500 milliard cubic metres. It was now clear that the Dutch had discovered a significant gas deposit, even though its full extent was not to become apparent until some time later.

As a result of yet more appraisal drillings and geological work, reserves estimates were reassessed and in October 1963 it was announced that they had risen to 1100 milliard cubic metres. Subsequently, further revisions were made with the result that by 1 January 1976 the proven reserves of Groningen stood at 1555 milliard cubic metres, after taking account of past production by that time of about 400 milliard cubic metres, or some 1715 milliard cubic metres if expectations are included. Moving on in time, there is now little doubt that the ultimate recoverable reserves of Groningen will probably turn out to be more than 2000 milliard cubic metres when the field has to be finally abandoned.

This puts the Groningen gas field among the top half a dozen of the largest known non-associated gas fields in the world. The actual place it occupies in the league table of the world's major natural gas accumulations is impossible to determine precisely. The reasons it is not possible to draw up such a table, which perhaps has little practical meaning in any event, is that various countries/companies have different standards and methods by which they measure so-called proven, probable and possible reserves. The composition, and hence calorific value of the gas in question, is another factor which may need to be taken into account when comparing one field with another if the actual useful energy content is an important criterion in such an assessment. Furthermore, it is only when a particular gas or oil reservoir is nearing the end of its useful productive life that it is possible to determine the full extent of the reservoir's ultimate recoverable reserves. In this connection, it will be appreciated that while production techniques are continually improving, many other factors of a financial, commercial and technological nature will have a bearing on when it is no longer economic or practicable to try to recover those hydrocarbons still remaining in the reservoir.

Leaving such considerations to one side, what is not in dispute is that in the Netherlands there had been found a major gas field by international standards, and moreover, this field was located onland, not only within an industrialised country with high living standards where gas could be readily utilised, but also within close proximity of the major industrial connurbations of Belgium, Germany and northern France—indeed within striking distance of southern Europe as well. The economic worth of and scope for Groningen gas was thus potentially considerably greater than for many similar large gas accumulations in various countries in the Middle East, Africa, in parts of the USSR, and elsewhere, which by geographic and geologic accident are more often than not remote from the world's main gas consumption markets. In this sense the Groningen gas field has virtually no peer, except perhaps the Panhandle/Hugoton field in Texas discovered in 1914 and now more than half depleted.

The Nature of the Groningen Field

Although such matters cannot be determined beyond dispute, it is now generally agreed that the Groningen gas field was not formed from the deposition of marine organisms many millions of years ago as was usually the case with many gas and oil reservoirs in other parts of the world. The probability in the case of Groningen was that the source of this natural gas was the Carboniferous strata underlying the gas-bearing formation—the Rotliegendes—from which the natural gas is now produced. It is believed that a later stage in the carbonisation of the coal layers in the Carboniferous formation must have caused the liberation of large volumes of methane, and that these must have migrated into the overlying red sandstone reservoir rock of the Rotliegendes; the thick layers of salt above the Rotliegendes obviously acted as a hermetic seal and prevented the natural gas from migrating further upwards and dispersing or escaping.

The Rotliegendes, an early Permian deposit, originated from a sand desert which was formed some 250 million years ago and covered a large part of northern Europe and what is now, in part, the North Sea. The Leman and West Sole gas fields, for example, in the UK sector of the southern North Sea are also located in the Rotliegendes strata.

Interesting though these facts may be to the geologist, the more crucial point of interest as far as this account is concerned is the composition of Groningen gas. Whereas natural gases produced from many gas fields in the world contain relatively small percentages of inert constituents such as nitrogen, carbon dioxide, etc., and have calorific values close to or frequently in excess of 1000 Btu per cubic foot (or say 9400 kilocalories per cubic metre), Groningen natural gas contains, inter alia, over 14 per cent nitrogen and has a resultant calorific value of barely 8400 kilocalories per cubic metre (kcal/m^3).

Table 5.1 gives the approximate composition in percentages by volume of Groningen gas and also the compositions for some other selected gas fields in the Netherlands and the North Sea. It should be noted that the data originate from specific analyses and changes can and do occur over the production life of a field.

TABLE 5.1 NATURAL GAS COMPOSITIONS

Component	*Groningen*	*Annerveen (Drente)*	*Dutch offshore field*	*Ekofisk (North Sea)*
Oxygen	<0.01	<0.01	—	—
Carbon dioxide	0.94	0.69	1.64	1.60
Nitrogen	14.27	4.17	1.93	0.52
Methane	81.28	90.53	91.95	85.93
Ethane	2.82	3.41	3.65	8.07
Propane	0.40	0.76	0.54	2.72
Butane	0.14	0.27	0.17	0.89
Pentane	0.03	0.08	0.04	0.21
Other hydrocarbons	0.11	0.08	0.08	0.06
Gross calorific value (kcal/m^3)	8 380	9 520	9 590	10 600

Within fairly narrow limits differences in the composition of natural gas can be accommodated in a gas distribution system without affecting adversely the satisfactory operation of consumers' appliances. However, beyond a certain range changes in gas quality become critical. The point to be made here is that an appliance, such as a domestic gas cooker or water heater, designed or converted to operate on Groningen gas quality will not necessarily operate satisfactorily or safely if it were to be supplied with a natural gas of a substantially different quality.

The decision was made fairly soon after the discovery of Groningen to distribute Groningen gas tel quel, apart, of course, from the necessary removal of entrained water and solids. This meant that in the conversion programme throughout the Netherlands, to be described later, appliances were converted to accept Groningen quality gas, a rather lower calorific value gas, because of its high nitrogen content, than some other natural gases which were later to become part of the overall Netherland's supply availability, albeit in minor proportions compared with the contribution made by Groningen gas. However, it is only in very recent years that the need for compatibility in introducing natural gases of varying qualities in the Netherlands has posed a challenge of any consequence, and so far this problem has been solved with customary Dutch efficiency without any upheaval in the market place.

This apparent digression from the main story is necessary if only to point up that not only was the Groningen discovery unique in its size and location, but that it was also rather different in its geological formation from most other major gas fields, and that at the same time it comprised a gas of a quality which had not been encountered on the same scale hitherto in, or close to, such a large prospective market.

Arrangements for the Exploitation of Groningen Gas

In searching for the most appropriate way in which the exploitation and marketing of Groningen gas should be organised, the government kept the following two principal objectives in mind: that the gas should be sold on a commercial basis consistent with the greatest possible benefit to the national economy; and that the gas should be introduced as smoothly as possible, without undue dislocation of the existing energy market.

After considerable discussions with all interested parties the government decided to end the existing public monopoly of bulk gas supply enjoyed by the State Gas Board, but to maintain nevertheless a large measure of direct state participation. In view of the fact that the wholly owned state entity, N.V. Nederlandse Staatsmijnen (Dutch State Mines, or more simply DSM), as the largest producers and distributors of solid fuels and coke-oven gas in the Netherlands would be affected by the introduction of Groningen natural gas into the energy market, the government decided to employ DSM as the instrument of state participation in the exploitation and marketing of this gas. In view of the considerable technical problems envisaged, it was also agreed that Shell and Esso, the joint shareholders of NAM, would play a large part not only in gas production, but also in the marketing sector, hitherto 100 per cent publicly owned, so that full benefit could be derived from their know-how, experience and financial resources. At the same time the Netherlands

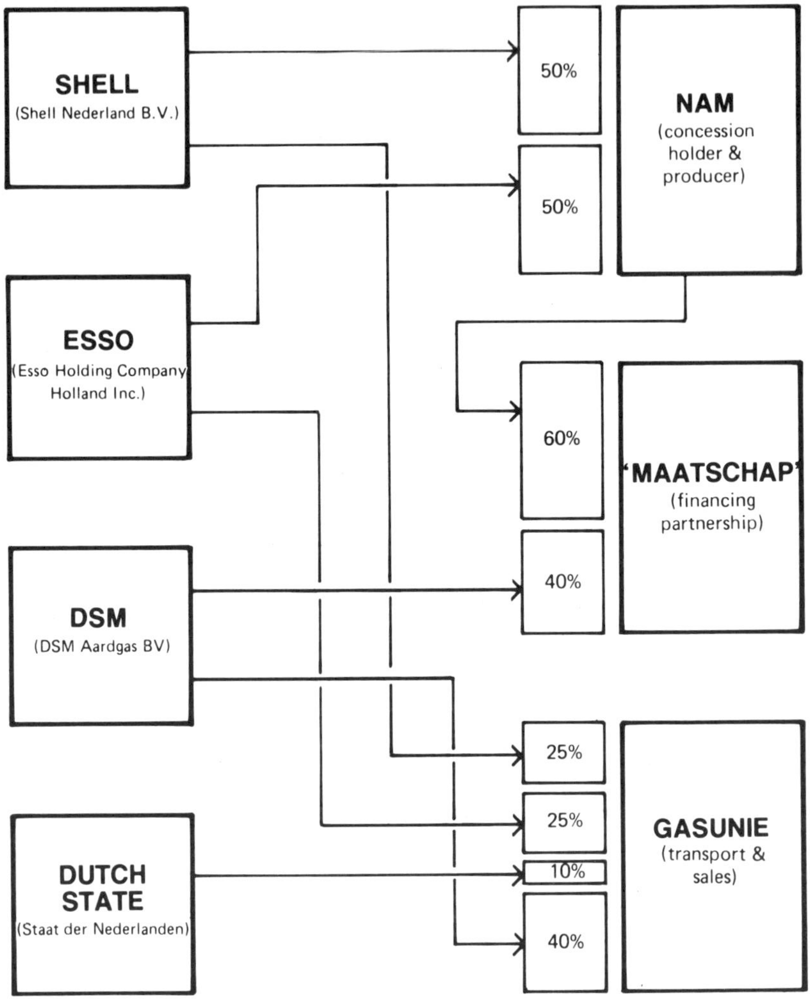

5.5 Corporate arrangements for Groningen gas

state, via DSM, would participate in the cost of and proceeds from the production of Groningen gas through a financing partnership, not a corporate entity in the normal sense, to be known simply as the 'Maatschap'.

For the transportation and marketing of gas within the Netherlands a new corporate entity was to be set up, N.V. Nederlandse Gasunie, in which Shell and Esso would each have a 25 per cent shareholding, DSM 40 per cent, and the Netherlands state a direct 10 per cent stake, giving the government an effective 50 per cent holding.

As far as exports were concerned, it was agreed between the government and NAM's shareholders that these were to be carried out by NAM for the account of Gasunie for which purpose a separate division of NAM would be set up which subsequently became known as NAM/Gas Export.

5.6 A typical Groningen 'cluster'. Seven wellheads can be seen in the background and the flare stack in the foreground (*N.V. Nederlandse Gasunie*)

The Agreement for Co-operation

The basis of the arrangements outlined above, as by then agreed between the principal parties concerned, was published by the Dutch Minister of Economic Affairs, de Pous, on 11 July 1962 in his 'Nota inzake het aardgas van Minister van Economische Zaken' (i.e. Natural Gas Note). On 4 October 1962, these proposals were passed without a division by the Dutch parliament and this paved the way for the setting up of the corporate entities necessary for its implementation. However, before this step was taken an 'Agreement for Co-operation' was signed on 27 March 1963, between DSM, Shell, Esso and NAM, to which the Minister affixed his signature on 4 April. Under this Agreement DSM had the right to nominate the managing director of Gasunie, and among other things it was also agreed that Gasunie would publish annual accounts.

On 6 April 1963, the 'Maatschap' was formed and N.V. Nederlandse Gasunie incorporated. In the following month NAM/Gas Export was formally established, and on 21 June of that year a fourth entity, Internationale Gas Transport Maatschappij N.V. (IGTM), was set up. The role of IGTM, owned as to 50 per cent each by Shell and Esso, was to provide technical advice and to develop gas transport and/or distribution ventures outside the Netherlands.

The organisational structure for the exploitation of Groningen gas was thus established. It differed fundamentally from what had gone before and in essence resulted in a 50/50 private industry/government partnership (Gasunie) in the transport and marketing of such gas within the Netherlands, and a 60/40 financing partnership (the 'Maatschap') in respect of production costs and proceeds for Groningen gas: the Maatschap arrangement did not apply to non-Groningen natural gas.

The remaining action necessary to set the wheels in motion was to grant to NAM the concession for the exploitation of hydrocarbons in the province of Groningen as part of the overall development plan. This was done on 1 May 1963. As regards any crude oil that might be discovered, it did not differ from earlier concessions in that NAM could exploit any such oil without state direction, subject to reimbursing its partners for costs attributable and to making over to government a royalty equivalent to 10 per cent of net profits.

The first delivery of Groningen gas by NAM to Gasunie took place on 9 December 1963.

Ministerial Discretions

In view of the importance of Groningen gas to the Dutch economy the Minister of Economic Affairs reserved to himself the following powers:

—the right of approval of the transfer price between the 'Maatschap' and Gasunie;

—the right of approval of the gas sales plan;
—the right of approval of conditions and tariffs for the delivery of gas by Gasunie to public distribution companies in the Netherlands, as well as the right to approve the price level for other customer categories;
—the right of approval for the construction of transmission lines and other equipment for the transportation and storage of gas;
—the right of approval of the Dutch border prices for export sales; and,
—the right to have a limited quantity of gas supplied by and for the account of Gasunie at prices and conditions as stipulated by him after consultation with Gasunie to customers to be designated by him, if he should consider this desirable to promote industrial development in certain parts of the country.

Although Gasunie was formed specifically for the purpose of bringing Groningen gas to market, it also had the right to purchase, transport and sell all refinery gas, coke-oven gas and natural gas produced in the Netherlands outside the Groningen concession in so far as such gases were supplied to third parties by the shareholders of Gasunie including their affiliated companies.

Construction of the Pipeline System

Once formed, Gasunie acquired the gas transmission system of the former State Gas Board of approximately 2100 kilometres (km), and that of the former Gasdistributiebedrijf (the gas distribution company of DSM) of about 800 km. The written-down book value of these assets at December 1963 was Dutch florins 162.3 million.

Before the formation of Gasunie, detailed studies had been made of a country-wide trunkline system, with the aim of making Groningen gas available almost everywhere in the Netherlands as soon as practicable so as to achieve quick penetration of the public distribution and industrial markets. The main trunklines to be built in the initial stage, i.e. during 1964, were:

	Pipe diameter		*Length*
	(inches)	*(mm)*	*(km)*
Hoogezand–Ommen–Zutphen	36	915	130
Ommen–Utrecht	36	915	110
Zutphen–Ravenstein–Boxtel	36	915	100
Revenstein–Geleen	24	610	110
Utrecht–Woerden	36	915	20
			470

A construction programme of this magnitude in one of the world's most densely populated countries was a major operation and presented a number of unique problems. For example, much of the cultivated land is owned by small farmers which necessitated the negotiation of a considerable number of rights of way. In fact the average number of landowners that had to be contacted for the 1964 construction programme was 12 per km, and in one part of the country it was as high as 35 per km. In total nearly 6000 right-of-way contracts had to be made. To facilitate matters the Minister of Economic Affairs established a Planologische Werk-commissie (Planning Working Committee) to serve as a final approval body for the proposed rights of way arrived at in consultation between Gasunie and the many public services involved.

Another major problem was water which caused three separate difficulties. First, was the crossing of many large rivers and canals, most of which had high levels of traffic which could not be interrupted for any length of time. Second, was the need to avoid any lasting interference to the country's immense and vital system of drainage and irrigation canals. In order to allow for possible future changes to and developments of this system, a minimum soil cover of 1 to 2 metres had to be provided over such pipeline water crossings. The third and greatest difficulty was the nature of the subsoil, combined with a very high water table. Much of the land is covered with only a few centimetres of top soil overlying sand or peat. Consequently, unless preventive measures were taken, as a pipe trench was dug it filled with water and the sides almost invariably collapsed. To combat this, and depending on local soil conditions, drainage wells had to be sunk up to 6 metres in depth at distances of from 1 to 2 metres apart. Water then had to be pumped out for many days before the water table had been sufficiently lowered and the sand had become dry enough to prevent the trench from caving in.

Construction work started in April 1964. By the end of that year 470 km of large-diameter, high-pressure pipe had been successfully laid. All this work was done in close consultation with the Dienst Voor Het Stoomwezen (National Bureau of Steam Engines) who ensured that appropriate standards of safety and construction were maintained.

By the end of 1977, Gasunie's main transmission system had grown to 3610 km, designed for a maximum pressure of 66.2 bar (960 psi). In addition, regional pipelines had been built of 6300 km, designed for a maximum pressure of 40 bar (580 psi).

Pipe-laying is seldom easy anywhere, but in the Netherlands this unglamorous and vital task was carried out with speed and commendable efficiency in exceptionally difficult conditions as also was the construction of metering and pressure regulation stations (14 in 1964 rising to 74 in 1977); blending, gas reduction and injection stations (18 by 1977); export terminals (19 by 1977); supply or city gate delivery stations (1104 by 1977); and not least of all, eight large compressor stations. This is not the end of the story and extensions to the pipeline grid and to other facilities are either under construction or planned for future years.

Conversion

Prior to 1964, which it will have been gathered was to all intents and purposes the take-off year for Groningen gas, about 380 000 connections (customers) out of a total of nearly 2.5 million were already suitable for or had been converted to receive natural gas. These customers were, of course, supplied with natural gas from sources other than Groningen. In step with the construction and completion of Gasunie's main transmission system and regional grids, it was necessary to convert over 2 million customers and about 5 million appliances from low calorific value manufactured gas to Groningen gas quality.

This task was given high priority. Conversion was the responsibility of the mainly municipally owned public distribution companies, but obviously their conversion efforts had to be undertaken in closest collaboration with Gasunie to ensure the minimum inconvenience and supply interruption to existing gas consumers. In 1964, nearly 100 000 customers were converted, in 1965 over half a million, in 1966 (the peak year) 740 000 customers, in 1967 570 00, and in the final year of 1968 about 180 000, i.e. a total of about 2.1 million over the five-year period of 1964 to 1968. In addition to the conversion of existing customers, new customers were being added to the grid all the time. For example, in 1967 new domestic connections alone amounted to about 100 000.

It is estimated that the total cost of the conversion programme, which was borne entirely by the public distribution companies, was some guilders 600 to 700 million, or an average of roughly guilders 250 per connection. In principle, the programme was so arranged that consumers whose appliances were suitable for conversion did not incur any additional expense. Consumers with old appliances which would have been very difficult to convert were offered very generous discounts on new ones and many consumers—as many as 40 to 50 per cent in some locations—took advantage of such offers to trade in their appliances, and/or to acquire additional appliances at attractive prices. Although this policy added to the distribution companies' immediate costs, it had the longer-term benefit of reducing future maintenance costs, improving safety standards by removing from the system many appliances which were inherently unsatisfactory, and where additional appliances were installed, increasing sales of gas per connection or household.

In spite of the large number of distribution entities involved, over 200 when natural gas was first introduced, but through some amalgamations and mergers this had come down to 170 by 1963 and to 137 by 1978, the conversion programme was accomplished with great success. In general, customer goodwill was enhanced rather than weakened.

The Reorganisation of the Public Distribution Market

In February 1963, the public distribution companies, already united in six regional organisations for the settlement of problems with regard to the supply of gas, decided to co-operate and founded the Commissie Samenwerkende Regionale Organen Gasvoorziening, known as SROG (Association of Regional Organisations of Gas Distributing Companies). SROG's principal object was to discuss and negotiate on behalf of the six regional organisations arrangements for gas supply and related conditions. By October 1963, negotiations between Gasunie and SROG resulted in an agreement on a price formula for the supply of natural gas to public distribution companies.

Gasunie and SROG also advised the public distribution companies on customer tariffs for natural gas with the aim of enabling these companies to achieve maximum sales penetration immediately conversion was completed. The approach adopted was to establish uniform consumer tariffs throughout the country, thus avoiding a possible multiplicity of different tariffs being created which might inhibit the rapid penetration of natural gas in some areas. At the same time agreement was reached with SROG that Gasunie would supply natural gas direct to industrial consumers with an annual consumption of approximately 2 million cubic metres or more. On 8 October 1972, SROG was reconstituted and renamed 'Vereniging van Exploitanten van Gasbedrijven in Nederland', otherwise known as VEGIN, the Netherlands Association of Gas Boards.

The Development of the Domestic Market

In the early 1960s some 87 per cent of all Dutch households relied on coal for home heating. The usual custom was to space heat one main room of the dwelling; in 96 per cent of these coal-heated households the appliance used was the flued stove. The remaining households at that time were heated about 9 per cent with oil and approximately 4 per cent by gas.

Against this background the availability of abundant supplies of natural gas provided far more scope for developing space and central heating than was ever feasible under an essentially manufactured gas regime. Competitive pricing was a further stimulus.

Initially, as sales of coal-fired space heaters fell away, oil-fired space heaters gained ground, but after 1964 sales of oil heaters also declined as natural gas became more widely available. The change was quite dramatic. At the beginning of 1965 sales of gas-fired heaters comprised about 70 per cent of total sales of space heaters, oil-fired heaters around 30 per cent, while sales of coal-fired heaters had dropped to barely 10 per cent compared with over 85 per cent at the start of 1962.

Central-heating installations, as distinct from space or room heaters, were not at that time particularly common. But again the advent of natural gas was to change this. Although precise figures are not available, about 10 per cent of all new houses (privately or municipally owned) in the early 1960s were equipped with central heating mainly either coal- or oil-fired. By the mid-1960s, half of the new housing was equipped with central heating of which 32 per cent was gas-fired and 18 per cent by oil or coal. And by the end of the 1960s these proportions had changed to over 70 per cent gas-fired, about 10 per cent heated by other fuels, with the balance not equipped with central heating. At the end of 1978 about 60 per cent of all Dutch homes were equipped with central heating of which about 94 per cent were gas-fired and 6 per cent oil-fired.

Over these years the average rate of completion of new dwellings was somewhat over 120 000 units per annum. The highest penetration of central heating was in multi-family houses, flats and apartments where block heating was usually installed. In these types of new dwellings natural gas achieved a saturation level approaching 100 per cent. While some of this growth in central heating, even in the absence of natural gas, would have taken place, given the rise in disposable incomes and the desire for higher comfort, much of the development was undoubtedly attributable to the strong sales appeal of natural gas founded on promotional tariffs and on the cheaper cost of gas equipment compared with oil and coal installations. The fact that gas required no storage space on consumers' premises, no ash disposal, and there was also its cleanliness and ease of control, were added advantages that the fuel enjoyed over its competitors.

The rapid penetration of gas-fired space and central heating resulted in the average consumption per domestic connection rising from less than 350 cubic metres in 1963, to over 1700 cubic metres in 1970, and to over 3500 cubic metres by the late 1970s. Actual average consumption rates per year are, of course, somewhat misleading because of the influence of temperature winter on winter, nevertheless the underlying trend is clear. Of these total consumption rates per connection, about 300 cubic metres per year were used for cooking and water heating. Consumption for these latter applications does not, of course, change to any noticeable extent over the years and is unaffected basically by seasonal influences and standards of living. Apart, therefore, from the acquisition of new customers, admittedly an important factor which cannot be entirely ignored, the main motor of growth for the domestic use of gas in the Netherlands was the heating market.

It is unlikely, other than in an exceptionally cold and prolonged winter, that average consumption rates will rise much further, indeed the trend will probably be in future for consumption rates to decline as has been the case in the United States in recent years. The main reason for this is conservation, that is to say improved home insulation. In this regard Gasunie has estimated that as much as 1000 cubic metres of gas per household per year could be saved in a normal winter if all gas-heated houses in the Netherlands were adequately insulated. If more stringent standards of insulation were applied nationally—and, of course, there comes a point when cost-effective considerations come into play—then perhaps even double this volume of gas could be saved. It remains to be seen to what extent insulation standards will be improved on a voluntary and/or government-inspired and subsidised basis in the coming years, but the balance of probability seems to be that some decline in average rates of consumption will undoubtedly occur.

Before leaving the domestic scene, available statistics indicate that natural gas in the Netherlands has achieved the highest rate of penetration of the heating market of any country in the world. The United States comes close to this, but bearing in mind the numbers of regions in that country where the need for space heating hardly arises, such comparisons are perhaps less meaningful than they may seem at first sight.

The Industrial Market

The arrangements agreed between Gasunie and SROG in 1963 resulted in a situation whereby supplies for small industrial consumers would be the responsibility of public distribution companies, while Gasunie would supply large industrial consumers—basically those customers whose annual consumption exceeded approximately 2 million cubic metres—direct. This review of the development of the industrial market concentrates on the latter category of customers from the time that Gasunie commenced operations (April 1963).

From modest beginnings, sales to industry, including power stations and chemical plants, became an increasing proportion of Gasunie's total deliveries of gas within the Netherlands. By 1969, they equalled in volume the off-take by public distribution companies, part of which in turn comprised indirect sales to small industrial consumers.

Initially the main outlets were chemical plants, where natural gas was used largely as a feedstock for the manufacture of fertilisers, the textile, ceramics, and food and drink industries. Sales of gas for power generation got underway in 1965, and with the introduction of interruptible-type supply contracts in 1967, expanded rapidly thereafter to the extent that barely two years later (i.e. as at end 1969), over 30 per cent of the total installed electric-generating capacity in the Netherlands was equipped to utilise natural gas. In the same year, of the estimated 530 or so industrial establishments with an assessed potential annual consumption of 2 million cubic metres or more, over 60 per cent had switched wholly or partly to natural gas firing.

As the industrial market grew so it was considered necessary to restructure the tariffication policy and to introduce a more uniform approach. To this end, on 1 January 1967, Gasunie adopted an industrial gas pricing system

5.7 Holland is famous for its greenhouses, most of which are now heated by natural gas (*N.V. Nederlandse Gasunie*)

which was linked to the price of heavy fuel oil. Tariffs were applied with fixed annual charges and variable commodity prices related to three ranges of annual consumption. The commodity or base price element of these tariffs was adjusted every quarter in accordance with a formula reflecting movements in oil prices. In this way it was possible to ensure that natural gas did not get unduly out of line for any length of time with competitive energy (i.e. oil) prices. While adjustments to the tariff system have been made over the years, the basic concept of indexing industrial gas prices to oil prices has been maintained since 1967.

The importance of the industrial market and how it has developed in relation to the public distribution sector is illustrated in *Table 5.2.*

TABLE 5.2 SALES OF NATURAL GAS BY GASUNIE WITHIN THE NETHERLANDS (MILLIARDS CUBIC METRES)

<table>
<tr><th>Market sector</th><th>1963</th><th>1965</th><th>1967</th><th>1969</th><th>1971</th><th>1973</th><th>1975</th><th>1977</th><th>1978</th></tr>
<tr><td>Power plants</td><td>nil</td><td>0.05</td><td>0.42</td><td>1.93</td><td>6.07</td><td>9.79</td><td>10.82</td><td>10.59</td><td>9.11</td></tr>
<tr><td>Chemical industry</td><td rowspan="2">0.22</td><td rowspan="2">0.51</td><td rowspan="2">2.19</td><td rowspan="2">5.16</td><td rowspan="2">8.80</td><td>2.96</td><td>2.89</td><td>3.23</td><td>3.15</td></tr>
<tr><td>Other industry</td><td>8.67</td><td>8.40</td><td>8.43</td><td>8.20</td></tr>
<tr><td>Total industry</td><td>0.22</td><td>0.56</td><td>2.61</td><td>7.09</td><td>14.87</td><td>21.42</td><td>22.11</td><td>22.25</td><td>20.46</td></tr>
<tr><td>PD companies</td><td>0.63</td><td>1.17</td><td>3.17</td><td>7.08</td><td>11.48</td><td>16.86</td><td>19.99</td><td>21.69</td><td>23.63</td></tr>
<tr><td>Total inland sales</td><td>0.85</td><td>1.73</td><td>5.78</td><td>14.17</td><td>26.35</td><td>38.28</td><td>42.10</td><td>43.94</td><td>44.09</td></tr>
</table>

Note: For some of the earlier years a detailed break-down is not available, and for some of the later years the split of sales between the chemical industry and other industry is approximate rather than precise.

Pricing Policies

Sales by public distribution companies to converted domestic consumers were initially made on the basis of the uniform tariff agreed between Gasunie and SROG in 1963. For space heating the additional costs for natural gas, i.e. over and above requirements for cooking and water heating, were about Dutch cents (D. cts) 12 to 14 per cubic metre depending on the amount of gas consumed. On an equivalent heat basis this compared with D. cts 11 to 16 per cubic metre for various qualities of coal, the main alternative fuel. Natural gas, bearing in mind its ease of use, was thus well able to compete in this particular sector of the market. For central heating the additional costs at, for example, a consumption of 3000 cubic metres per year were about D. cts 11 per cubic metre. This was just about competitive with domestic heating oil, given the price advantage of a gas boiler over an oil boiler.

Oil started to fight back almost immediately by reducing its prices for domestic (central-heating) oil down to a level equivalent to about D. cts 8 per cubic metre by the end of 1964. The gas industry reacted to this competitive pressure and in February 1965 lowered its selling prices for customers consuming in the range 2100 to 20 000 cubic metres per year. For such customers the tariff was amended to a fixed annual charge of guilders 150 and a commodity charge of D. cts 7 per cubic metre. As these changes affected adversely the profitability of the public distribution companies, the price they paid Gasunie at the city gate for their supplies was revised to D. cts 5.8 per cubic metre. This price enabled the distribution companies to achieve a very high penetration of the heating market within a few years and to acquire many new customers which were not hitherto connected to the gas grid.

Thereafter the basic approach to consumer gas prices in the heating market was to set tariffs at levels ostensibly competitive with oil prices, also to relate reducing unit costs with increasing consumption in order to maximise individual off-takes. In this way large average consumption rates per customer were achieved which helped public distribution companies to reduce the unit cost of gas supplied, to increase their total revenues, and hence to recover the cost of conversion relatively quickly. In practice gas tariffs were set generally rather lower than their true competitive or market value, which can be regarded as the equivalent price of gas oil plus a premium for such factors as convenience of use, flexibility of control, cleanliness, etc.

Over the years prices have, of course, changed in line with the changes in the general economic and energy environment. By 1976, the price had risen to a little below D. cts 23, compared with an equivalent oil price of D. cts 29 per cubic metre. On 1 January 1979, the price was increased to D. cts 25 per cubic metre for customers consuming more than 600 cubic metres per annum, still somewhat below the equivalent oil price.

Throughout the period reviewed, Dutch gas prices for the domestic consumer have been almost invariably competitive with oil prices and among the lowest gas prices in Western Europe. The pricing policy followed has been remarkably successful to the point where since the early 1970s some 80 to 90 per cent of the new housing stock has been equipped with gas-fired heating. Similarly, average annual consumption rates have risen dramatically from about 300 cubic metres in 1963 to between 3500 and 4000 cubic metres by the late 1970s, the actual rates, of course, depending in part on the severity of the winter.

The initial approach to pricing for the industrial market was of necessity much more tentative than that for the domestic market. However, since September 1964, when the first industrial tariffs were approved, a number of changes have been made with the aim of introducing the maximum flexibility possible so as to cater for the specific requirements of a wide variety of types and sizes of industrial consumers. From the start industrial tariffs have been linked to the cost of heavy fuel oil, gas's main competitor.

As already mentioned large industrial consumers were supplied direct by Gasunie, but for smaller industrial consumers public distribution companies were granted a margin so that they could apply Gasunie's industrial tariffs and maintain a uniform pricing structure. In view of the complexity and variations in the tariffs, space precludes any detailed account of the actual tariffs imposed and their subsequent changes. However, as a very broad indication one can say that through to about the early 1970s actual prices were within about 20 to 25 per cent either side of the equivalent cost of fuel oil. Since that time for new contracts, or for old expiring contracts coming up for renewal, the general approach has not only been to secure full parity with fuel oil, but also to charge a premium and to ensure that gas prices are appropriately indexed with movements in competitive oil prices.

Exports

As discussed previously, the framework arrangements for the export of Dutch gas were laid down in the Minister of Economic Affair's 'Natural Gas Note' to Parliament of July 1962 and the necessary corporate structure to undertake this was established in the following year.

However, prior to 1963 the State Gas Board had already signed a contract for the export of small quantities of Dutch natural gas to the German company, Energieversorgung Weser-Ems A.G. (EWE). Initially natural gas was supplied from the Schoonebeek concession, but in 1963 the supply source was switched to Groningen. In that year exports of Groningen gas to this company amounted to some 5 million cubic metres; by 1970 they had increased to 191 million cubic metres—a modest beginning.

During the early/mid-1960s a series of export deals were concluded by NAM/Gas Export with Thyssengas A.G., Ruhrgas A.G., and Deutsche Erdgas Transport G.m.b.H. (DETG) in Germany, Distrigaz S.A. in Belgium, and with Gaz de France. Concurrently, Shell and Esso acquired minority shareholding participations in Thyssengas (25 per cent each), in Ruhrgas (14.7445 per cent each, including indirect interests via affiliated companies), and in Distrigaz ($16\frac{2}{3}$ per cent each); subsequently, on 5 May 1975, Esso sold its shareholding in Distrigaz to Belgian interests.

In order to implement these export deals new gas transmission pipeline systems were built in Belgium, both for the local market and for the transport of Dutch gas to the French border for Gaz de France. In addition, three pipeline companies were formed in Germany: Nordrheinische Erdgastransport G.m.b.H., to transport gas from the Dutch border to the Bergisch Gladbach area for Ruhrgas, Thyssengas and DETG; Mittelrheinische Erdgastransport G.m.b.H., to transport gas from Bergisch Gladbach to Russelsheim for Ruhrgas and DETG; and Suddeutsche Erdgas-Transport G.m.b.H. to transport gas from Russelsheim to Lampertheim for Ruhrgas and DETG.

The magnitude of these commercial and pipeline construction activities, and not least of all the related conversion work that had to be undertaken in the export markets, can best be judged from *Table 5.3* which shows how exports developed over the period 1963 to 1970; sales to the local Dutch market are also shown for comparison purposes.

TABLE 5.3 EXPORTS OF DUTCH GAS: 1963–70 (MILLIONS CUBIC METRES)

Customer company	*1963*	*1964*	*1965*	*1966*	*1967*	*1968*	*1969*	*1970*	*Cumulative 1963–70*
EWE	5	10	37	45	69	124	156	191	637
Thyssengas	—	—	—	3	118	516	1 011	1 824	3 472
Ruhrgas	—	—	—	—	155	895	1 415	1 491	3 956
DETG	—	—	—	—	—	6	84	240	330
Germany	5	10	37	48	342	1 541	2 666	3 746	8 395
Distrigaz	—	—	—	87	504	1 329	2 811	4 493	9 224
Gaz de France	—	—	—	—	271	1 487	2 084	3 032	6 874
TOTAL EXPORTS	5	10	37	135	1 117	4 357	7 561	11 271	24 493
Netherlands	806	1 001	1 727	3 248	5 781	9 608	14 172	20 136	56 479
TOTAL SALES	811	1 011	1 764	3 383	6 898	13 965	21 733	31 407	80 972

In the second half of the 1960s/early 1970s further export contracts were concluded with SNAM S.p.A., a subsidiary of Ente Nazionale Idrocarburi (ENI), the Italian state oil and gas company, and with various other German companies: Rheinisch-Westfälisches Elektrizitätswerk A.G., Vereinigte Elekrizitätswerke Westfalen A.G., Gas-Union G.m.b.H., and Gewerkschaft Brigitta. For the supply of Dutch gas to Italy a new main trunkline was built from the Dutch border through West Germany and Switzerland to northern Italy. One of the arrangements for this particular contract included provision for a small part of these supplies to be delivered en route to Swiss gas interests.

An export prospect which did not materialise, however, was the possibility of delivering Dutch gas by submarine pipeline to the United Kingdom. This was considered by Dutch and British interests in 1964 and in 1965, but with the discovery of the West Sole field in October 1965, followed by more discoveries in the UK sector of the North Sea in 1966 and in subsequent years, the need for the UK to import natural gas from the Netherlands first receded and then disappeared.

The development of Dutch gas exports since 1970, including again for comparison purposes sales to the local market as well, is as shown in *Table 5.4*:

TABLE 5.4 EXPORTS OF DUTCH GAS: 1971–78 (MILLIARDS CUBIC METRES)

Customer country	*1971*	*1972*	*1973*	*1974*	*1975*	*1976*	*1977*	*1978*	*Cumulative 1971–78*
West Germany	6.5	9.8	13.8	18.4	21.0	21.8	22.7	17.6	131.6
Belgium	6.2	7.9	9.7	11.6	11.3	12.0	11.1	10.2	80.0
France	4.4	6.1	7.9	9.5	9.4	10.2	9.7	10.0	67.2
Italy	—	—	—	1.7	4.5	6.1	6.0	6.8	25.1
Switzerland	—	—	—	0.2	0.5	0.5	0.5	0.5	2.2
TOTAL EXPORTS	17.1	23.8	31.4	41.4	46.7	50.6	50.0	45.1	306.1
Netherlands	26.4	34.1	38.3	41.1	42.1	43.7	43.9	44.3	313.9
TOTAL SALES	43.5	57.9	69.7	82.5	88.8	94.3	93.9	89.4	620.0

The first year when total exports (41 405 million cubic metres) exceeded total sales of natural gas within the Netherlands (41 113 million cubic metres) was 1974. Since that time annual exports have continued to comprise variously from 50.4 to 53.7 per cent of combined Dutch gas sales for internal and export consumption.

As we have moved ahead in our story, let us retrace our steps a little in time. On 1 April 1975, Gasunie took over the activities of NAM/Gas Export in respect to the export of Dutch gas on the same contractual terms and conditions as were then applicable to such sales. The activities of IGTM, which at an earlier date had been merged with those of NAM/Gas Export, resulted in a decision to disband IGTM as a corporate entity in 1971.

It is not possible to record the various prices and related contractual conditions applicable over the years for Dutch gas exports as these are not published and remain private matters between the parties concerned. However, it is possible to state that most of the natural gas export contracts that were concluded in the 1960s were at fixed border prices, in a period when energy prices were falling and when energy as a whole was in plentiful supply. These contracts also stipulated that periodic price adjustments could be negotiated. Such a condition is what one might expect for contracts which originally had a life span of 20 to 25 years. It would not have been reasonable to expect either contracting party to have remained rigidly bound in every single detail to a contract over such a period given the unforeseeable changes in external conditions that could and did arise. For example, during and immediately after the oil price rise shocks of 1973 and 1974, the previous system of more or less fixed Dutch border prices was abandoned and more flexible pricing mechanisms linked to oil prices were introduced by common agreement. Similarly, and somewhat later in time, as many export customers began to receive increasing volumes of natural gas from other sources, which for various reasons had to be accommodated by the purchasing companies concerned at high load factors, further changes in the supply conditions of Dutch gas to these export customer companies became operationally desirable. Accordingly, in the next round of renegotiations which started in October 1976 and extended over the next year to 18 months, export border prices were readjusted in exchange for greater off-take flexibility, coupled with extensions to the contract period, but with no increase in the total volumes to be supplied over the revised and extended contract period.

As a consequence of these latter adjustments exports in 1978 were somewhat lower than originally planned and indeed were down on the level achieved in 1977. However, as already indicated, exports will now continue for a longer period of time and will terminate less abruptly, a situation sought and welcomed by the export customers concerned.

The depressed state of economic activity in Europe, the mild winter weather conditions that have generally prevailed (apart from the second half of the winter of 1978/79), energy conservation policies, a gradual reduction in the use of natural gas for power generation purposes (at governmental request or direction), and the growing availability of gas supplies from non-Dutch sources, have all combined to somewhat dampen the rate of increase in Dutch gas exports in recent years. Nevertheless, the level reached of some 45 to 50 milliard cubic metres per annum since 1975 has been a remarkable achievement by any standards. In the absence of any very large new discoveries onland or offshore the Netherlands, it is improbable that any new export deals will be made. Remaining recoverable reserves of Dutch gas are still considerable—approximately 1680 milliard cubic metres at the beginning of

1978—but after allowing for existing export commitments, the balance available, plus any new discoveries and contracted-for imports (about 140 milliard cubic metres at present), can be expected to be retained for ultimate use within the Netherlands.

Exports of Dutch gas have not only represented a substantial source of foreign exchange earnings for the Dutch economy, but have also reduced the out-goings for oil imports. Moreover, Dutch gas was until September 1977 the only source of natural gas supply for Belgium, is still the largest single component of supply for the German gas market, is the predominant source of gas supply for northern France, and more recently has built up to become an essential part of Italy's supply portfolio. Without Dutch gas some of these European gas markets would have had a tough fight to survive, especially in the 1960s, in the face of severe competition from cheap and abundant oil supplies at that time. Not all of the gas exported from the Netherlands has come from Groningen, but it was the discovery of the Groningen gas field above all else that enabled exports on this scale to be contemplated and realised.

Two of perhaps several yardsticks can be used to illustrate the importance of Dutch gas. First, Dutch gas as a whole, i.e. including exports and non-Groningen gas supplies, has constituted nearly half of the total consumption of natural gas in continental Europe during recent years. Second, exports of Dutch gas alone represent (at say the current rate of off-take of around 45 milliard cubic metres per annum) the equivalent of over 700 000 barrels per day of oil—energy which otherwise would most probably have had to be imported from non-European sources. If local consumption in the Netherlands is included, the figure then rises to an impressive 1.5 million barrels per day of oil.

By end 1978, a total of about 330 milliard cubic metres of Dutch gas had been exported, compared with some 370 milliard cubic metres consumed within the Netherlands. For current export contracts, supplies delivered to date represent somewhat over one-third of the total contracted-for volumes which (including past deliveries) amount to a total of about 925 milliard cubic metres. Under present arrangements the last export deliveries are due to terminate in the late 1990s/year 2000, and from then on, unless some dramatic and unforeseen events occur, those countries and companies which have hitherto relied exclusively or to a very large extent on Dutch gas supplies will need to have taken timely measures to ensure that supplies from non-Dutch sources are by then available in sufficient quantity to meet their demands for gas in the market place. One perhaps needs hardly add that the companies concerned are fully conscious of this situation and have already or are about to take appropriate steps to safeguard their own interests and those of their customers.

Liquefied Natural Gas

As sales of gas expanded, particularly for space-heating purposes, so the range between average consumption in the summer and peak off-take demand in the winter sharpened. In order to smooth out these fluctuations to some extent, Gasunie had adopted the common industry practice of contracting part of its supplies for certain large consumers, notably power stations, on an interruptible basis. In the Netherlands, such interruption comes into effect as soon as the average 24-hour ambient temperature drops below zero, which generally speaking occurs for about 30 days during a typical Dutch winter.

Gas thus saved is then available for the public distribution market which is not subject to interruption. As a design feature of its transmission system, Gasunie laid down that sufficient capacity must be available to meet the non-interruptible demand at an average 24-hour temperature of minus 15°C, even though temperatures of this severity are exceptional and indeed temperatures normally seldom go below about minus 10°C.

Apart from the in-built supply capacity and flexibility already available, in 1974 Gasunie decided to supplement this with a liquefied natural gas (LNG) peak-shaving plant in the more southerly part of their system to cope with any very high demands that could arise in that area. It was further decided to build this facility on the Maasvlakte, an area reclaimed from the sea, close to the densely populated industrial city and seaport of Rotterdam; construction of the installation was started in 1975.

Because of the high nitrogen content of Groningen gas it was doubted if Groningen gas could be liquefied, stored, and subsequently regasified satisfactorily as a homogenous mixture. The probability would be that the difference in boiling points of nitrogen and methane/ethane would result in a varying gas quality when the LNG was regasified. To extract and ventilate the nitrogen during liquefaction would also cause quality problems when the predominantly methane content LNG had to be fed back into the distribution system. The solution picked on, therefore, was to store methane and nitrogen separately after liquefaction so that after evaporation they could be mixed again in the same proportions as the original feed gas.

The actual plant built on the Maasvlakte comprised basically much the same type of liquefaction and evaporation facilities as described in Chapter 7, together with two storage tanks for liquefied methane/ethane each of a capacity of 57 250 cubic metres, and one tank of 19 000 cubic metres for the storage of liquid nitrogen. The inner LNG tanks were made of 9 per cent nickel steel, and the outer tanks of carbon steel with more than 1 metre of Perlite (granulated vulcanic) insulation between them. Strict safety regulations were imposed with the result that what can best be described as pre-stressed concrete 'containers' (at a cost of over guilders 30 million) were also built at a distance of 1 metre around and below the outer carbon-steel tanks to house each of the two LNG tanks. And in order to ensure that these 'containers' were not perforated, all necessary piping to the LNG tanks was fed through the top. A yet further precaution was to surround each tank site with a bund wall, in addition, of course, to the usual thermo detectors and fire-fighting equipment.

The Maasvlakte LNG peak-shaving plant, which cost about guilders 160 million, was officially opened on 12 May 1977.

The second aspect of Gasunie's LNG activity has yet to be consummated in practical terms, nevertheless as it is already under development it is not inappropriate to accord it brief mention in this historical account. In order to supplement its longer-term supply availability, Gasunie, in association with German gas interests, entered into a contract with Sonatrach, the Algerian state oil and gas entity, on 30 June 1977, for the importation into the Netherlands of approximately 5.6 milliard cubic metres of LNG per annum over 20 years commencing in 1983/84.

After considerable discussion the proposal that the site for the import terminal for such supplies should be the Maasvlakte/Rotterdam area was not agreed to and instead a new location of Eemshaven in the Eems estuary, close to the Groningen gas field, was selected by the Dutch government in 1978. Implementation of this contract and the construction of the necessary reception facilities at Eemshaven are matters for resolution in the coming years.

General Conclusions on Progress to Date

As stated at the beginning of this chapter, this has been largely the story of the development of Dutch natural gas since the discovery of the Groningen gas field in 1959. Groningen gas, supplemented by other natural gas discoveries onland and offshore the Netherlands, has had a profound effect not only on the Netherlands itself, but also on the gas industries in several other European countries. For the Netherlands it has resulted, inter alia, in the guilder becoming one of the world's stronger currencies at a time when many other currencies have weakened because of various factors such as inflation, sluggish economic activity, rapid increases in the prices of imported oil, industrial unrest, social changes and so on. The Netherlands has not been without some of these problems, but it had the advantage over most other industrialised countries in the western world of finding and developing a major energy resource which was more than sufficient for its own internal long-term requirements: by chance, and more or less at the right time, it found itself with a much wanted and readily exportable commodity of consequence. Taking one year as an example, which is not necessarily typical but is nevertheless illustrative, it is estimated that in 1977 the earnings from gas exports, together with the reduced need for imported oil, resulted in a net benefit to the Dutch national economy of some guilders 14 000 million—an impressive sum for a relatively small country.

Most informed observers would argue with conviction that the blend of central and municipal government and private industry has been just about right, or at least has been successful in ensuring that Dutch gas was developed in an expeditious way for the common good. While some critics might challenge this, surely even they cannot dispute that the Dutch developed this particular natural energy resource in the most competent, economic, commercial and technical fashion. Certainly, if one puts to one side the views of the coal, oil and other competing energy industries, from the gas industry standpoint the development of Dutch natural gas in the 1960s and 1970s must be rated as one of the world's success stories.

Today, nearly 400 major industries in the Netherlands use natural gas. Over 90 per cent of all Dutch homes are connected to the gas supply grid, and in turn again over 90 per cent of these gas-connected homes are heated with gas. In terms of primary energy consumption, natural gas now supplies approximately 52 per cent of the Netherlands' total requirement, a percentage that has not been matched in any other industrialised country in the world to date. Inevitably this percentage will decline slowly over the remainder of this century as the resource is finite and thus not inexhaustible, while at the same time population and industrial growth, improved purchasing power and living standards and similar factors, tend to drive up the need for more energy inexorably. Conservation in its widest sense, i.e. the more effective use of energy, may moderate the average rate of increase in energy consumption, but it is likely to be many years before the overall trend is halted and subsequently reversed. However, for certain market sectors, for example the domestic heating market, the effects of conservation are likely to be felt much sooner than, say, the use of gasoline for automotive purposes.

The Dutch have been very conscious of their finite and depleting natural gas resources for some years past. In fact every year since 1963, Gasunie has submitted a Plan van Gasafzet (Gas Marketing Plan) to the Minister of Economic Affairs. Since 1975 summaries of these plans have been made available to a much wider public audience. In essence these plans examine the supply and demand prospects for Dutch gas over both the medium term (next 5 years) and the longer term (up to 25 years ahead). Policies and strategies are suggested as to how available gas resources might be best utilised to meet known and foreseeable requirements against the background of available resources and likely additions, including the need for the acquisition of gas supplies from non-Dutch sources.

Apart from North America, this public discussion of how and in what manner gas may fit into the Netherlands' future pattern of energy needs is a relatively new innovation, although in all fairness it must be stated that several other countries have now started to open up this subject, indeed the whole range of prospects for all forms of energy, to much wider debate than was ever the case in the past. Nevertheless, it is not untypical for the Dutch to be among the forefront of this participative and consultative type of approach which accords with the overall changes now taking place in society around the world.

Some readers of this account may feel that the author has been too eulogistic in this review of the development of Dutch gas. Undoubtedly with hindsight some things might have been done differently with perhaps possibly better results. But no situation is either ideal or static; on balance it would seem that mostly the correct decisions were made on the basis of the known circumstances of the time and the best possible prognosis of the future that could be made when such decisions were taken.

Some of the more important lessons that can be drawn from the development of Dutch gas, in the way it evolved, are that the interests and objectives of both central and local government and private industry are not necessarily in conflict, but can be harmonised and be mutually beneficial. Second, it is possible to achieve a satisfactory working relationship between such disparate interests which does not in turn detract from the overall public interest. Third, and on a wider plane, without Dutch gas, and Groningen gas in particular, the gas industry in north-west Europe would probably be by now in a rather sorry plight. Fourth, the absence of cumbersome and time-consuming regulations and procedures, but with appropriate governmental involvement and powers of sanction, has enabled Dutch gas to make a satisfactory, timely and important contribution to Europe's energy needs, especially in the post-1973/74 period when all the upheavals on the hitherto fairly predictable oil scene took place.

These generalised observations are not to say that the way in which Dutch gas has been developed would have been necessarily suitable and applicable to other countries, but simply that in this particular environment it appears to have fitted the bill very well. In contradistinction, many would consider that the quite different regulatory approach followed in the United States was eminently suitable for the circumstances of that market, at least in the pre-1970 era. In yet other countries the conduct of the gas business ranges over the whole spectrum from a total government monopoly to a business which is essentially in the hands of private industry with little, if any, direct governmental intervention. Future historians will need to consider and review whether the conduct of the Dutch gas business was handled with appropriate perception and adroitness in the 1980s and 1990s. Be that as it may, evidence to date suggests that the way the Dutch have gone about the revitalisation of their gas industry so far has been in the main a praiseworthy achievement.

6: The Union of Soviet Socialist Republics

General Considerations

Generally speaking, information on the Soviet gas industry is not as extensive or consistent as is the case for most Western countries. Such data that are available originate principally from governmental and other official sources. There is little scope for verifying and adding to such data from independent sources, which in the main tend to recycle or interpolate official data and published reports with varying degrees of accuracy.

In spite of these reservations, there is no dispute that the USSR is today the world's second largest producer and consumer of natural gas. Equally, few geologists would argue that not only has the Soviet Union the largest proven gas reserves of any country in the world, however these may be defined, but that it also enjoys the prospect of discovering extremely large quantities of additional gas in the years ahead.

Wherever possible the most recent published statistics and related information have been used, but for the above mentioned reasons certain figures that have been quoted may not necessarily be as precise as one would wish. Nevertheless, in spite of the doubts that may exist as to their accuracy, they serve to indicate the underlying trends and general orders of magnitude applicable to the USSR's gas industry.

While the main thrust of this chapter is on the post-Second World War period when the development of the USSR's enormous natural gas potential really got underway, a brief resume is included—to the extent that records are available—on earlier developments and events.

Finally, for those readers who may not be too familiar with the Soviet scene, related terminology and abbreviations, the following may be helpful. The Union of Soviet Socialist Republics (the USSR) consists of 15 Soviet Socialist Republics (SSR) of which the Russian Soviet Federative Socialist Republic

(RSFSR) is the largest. An Autonomous Soviet Socialist Republic (ASSR) is an administrative area giving some political recognition to minority nationalities: there are 20 ASSRs all subordinate to the SSR in which they are situated. A Krai is an administrative subdivision subordinate to the SSR in which it is situated. An Oblast is also an administrative subdivision. However, many Oblasts are larger in area than the smaller SSRs, but they are always subordinate to the SSR in which they are situated.

Generally speaking, references to Russia and Russian in this chapter refer to pre-1917 days, while references to the USSR and Soviet (Union) concern the post-1917 period, thus drawing a distinction between the Czarist and Communist eras.

Pre-Revolutionary Days

One important event in Russia's history, which was ultimately to have a significant effect on the gas industry, was the expansion of Russia beyond the Urals into northern Asia. This started in the reign of Ivan IV (the Terrible) and resulted in Siberia being finally annexed to Russia in 1699 in the reign of Peter the Great. Two and a half centuries later Siberia was to become Russia's principal source of natural gas reserves and is now probably the world's most prolific gas-bearing region so far discovered.

Evidence of the date of the first commercial use of manufactured gas in Russia has so far eluded the author. In all probability it was somewhere around 1850, or possibly a few years earlier; certainly by the 1860s manufactured gas was being distributed in Moscow, St Petersburg, Kharkov, Rostov, Riga and Odessa for street-lighting purposes. By 1867, some 6000 street lights were lit by gas in Moscow. Around the turn of the century gas was also being used for household purposes, albeit on a limited scale. Just before the First World War total manufactured gas supplies had reached the relatively modest level of about 17 million cubic metres per annum. In 1914, about 3000 households were using gas in both Moscow and St Petersburg, but the extent to which gas was being utilised elsewhere in Russia is unclear.

Turning now to natural gas, Russia, like several other countries, also has its accounts of Eternal Fires. Perhaps the most famous of these concerns the celebrated Temple of Zoroaster, near Baku, which for over 2500 years was visited by the Guebers, or fire-worshippers of Asia. Interestingly, the temple contained a crematorium where the bodies of faithful devotees were consumed by the sacred fires of natural gas—an application for gas which exists today! Of more relevance is that Baku, where the first oil well was drilled in June 1871, was also most probably the first place in Russia where natural gas was put to practical use in modern times. Mr W. F. Hume records in a visit he made to Baku in 1896 that: 'natural gas is made use of for lime-burning, for every domestic purpose, and as a fuel for boilers; it is only necessary to sink a pipe a

6.1 An example of sales promotion in the early days of manufactured gas

few feet into the soil to obtain on ignition a flame of considerable length.'

While it is not possible to determine how much natural gas was utilised in these early days, all available accounts of these times point up that substantial amounts of natural gas produced in association with crude oil were vented and flared, gas utilisation being limited to very small quantities at or close to the points of oil production—a situation, of course, not dissimilar to North American experience of those days.

The second major historical event which affected the whole political and economic structure of the country was, of course, the October Revolution of 1917. This led in time to a total state-planned economy and the institution of five-year plans in 1928. Before these were introduced, and after the disturbances of the civil war had settled down, utilisation of natural gas by the mid-1920s had built up to the equivalent of about 130 million cubic metres per annum.

Soviet Plans and Developments between 1928 and 1945

The First Five Year Plan (1928–32) was concerned chiefly with developing heavy industry and power generation capacity, modernising the country's transportation system, and the collectivisation of agriculture. The development of both manufactured gas and natural gas received scant attention from the central planning authorities; the country had more pressing tasks on its hands. In spite of a steady increase in output the gas industry continued to play a minor role in the country's energy economy. Attempts were made to make better use of available gas production, nevertheless the flaring and venting of gas was still a matter for concern.

In 1928, consumption of natural gas amounted to 304 million cubic metres, and by 1932 this had built up to a little over 1000 million cubic metres, constituting about 1.3 per cent of total USSR primary energy consumption. How much additional gas was vented and flared is not known.

The Second Five Year Plan (1933–37) saw the completion of the collectivisation of agriculture and continued improvements in the transport system. Industrial development and the modernisation of the armed forces received high priority, whereas the development of natural gas proceeded slowly, still being confined very largely as in past years to the use of associated natural gas close to the points of production.

The principal objectives of the Third Five Year Plan (1938–42), which were to develop the oil, coal, lumber and chemicals industries, as well as the mechanisation of agriculture, further improvements in transportation, etc., were interrupted by the outbreak of the German-Russo war on 22 June 1941. In the preceding year (1940), natural gas consumption had risen to 3219 million cubic metres of which nearly 90 per cent was associated gas largely from oil fields in the western Ukraine. This was supplemented by some 200 million cubic metres of manufactured gas. Natural gas's share of total primary energy consumption was approximately 1.8 per cent.

Output and consumption of both natural and manufactured gases declined during the war—although the actual volumes are not known—but it is recorded that they regained their pre-war 1940 levels in 1946. This was the year in which the first long-distance natural gas pipeline was commissioned—a 12 inch diameter line of 788 km from Saratov to Moscow.

Apart from the interruption and upheaval of the war, these five-year plans had been an undoubted success in expanding and modernising existing industries, developing new industries which, coupled with an improved transport system, were to provide the foundation of the Soviet Union's future economic strength. Whereas in all pre-war plans no significant effort had been allocated to the development of the manufactured gas industry, or to the exploration for and utilisation of natural gas, in the post-war plans this was changed. First, emphasis was given to manufactured gas, and then when this effort proved disappointing, to natural gas.

The Fourth and Fifth Soviet Plans

The main task of the Fourth Five-Year Plan (1946–50) was inevitably the restoration of the damage wrought by the war. Low priority was given to oil and natural gas production, with much larger investments being made on synthetic liquid fuel plants, the production of gas from coke, lignite and shale, and the underground gasification of coal. The Fourth Plan envisaged a considerable increase in manufactured gas and set a target of 1800 million cubic metres out of a total forecast gas supply in 1950 of 6800 million cubic metres per annum.

Work had already started in August 1945 on the world's first commercial size gas from shale plant at Kokhtle-Jarve in Estonian SSR. This was said to be the largest single investment project under the Fourth Plan and was brought on-stream in April 1949. A 200 km pipeline was built from Kokhtle-Jarve to Leningrad in the preceeding year and in due course some 200 000 households in Leningrad were supplied regularly with gas made from shale. In 1953 a second 140 km pipeline was completed connecting this plant with Tallinn, the Estonian SSR capital, resulting in a further 55 000 households being supplied with shale gas.

In 1942, during the previous Third Plan, Podzemgas, the Soviet entity responsible for gasification projects, completed the world's first underground coal gasification plant near Moscow. The initial cost of gas from this plant was roubles 185 per 1000 cubic metres, but as the plant was expanded and productivity improved, this cost came down over the years to roubles 37 per 1000 cubic metres. During the Fourth and Fifth Plans further coal gasification plants were built at Donbass, southern Ukraine (in 1948), and in the mid-1950s in the Kuzbas region of central Siberia, at Shatsk near Moscow, and at Angren in central Asia. The Shatsk plant was designed to produce 660 million cubic metres of coal gas per annum using nearly one-quarter of a million tons of coal, while the capacity of the Angren plant was 2300 million cubic metres of gas and 650 000 tons of brown coal per annum.

The basic concept of these plants comprised the passing of a stream of air through subterranean channels cut by hand or by machine in the coal seams, and then igniting this air stream by remote-controlled devices. The quality of gas produced was very poor, approximately 1000 kilocalories per cubic metre, about one-quarter of the calorific value of coke-oven gas or one-tenth of that of natural gas. Because of its low calorific value the uses for such gas were limited mainly to low-grade, under-boiler industrial applications, in particular for steam raising in power stations located near to the point of production as the long-distance transmission of such gas was uneconomic.

The above mentioned target of 1800 million cubic metres of manufactured gas in 1950 was not realised. In fact output of manufactured gas rose from 100 million cubic metres in 1945 to only 420 million cubic metres in 1950, and it was not until eight years later that the Fourth Plan target of 1800 million cubic metres was reached.

Although the Fifth Five-Year Plan (1951–55) was less enthusiastic on gasification processes, provisions were made to continue investments in projects already in hand or at an advanced stage of planning and design. Between 1945 and 1958 about roubles 150 million were spent on manufacturing gas from coal and shale and over roubles 40 million on underground coal gasification. But in relation to the effort and investment made over the period, the results achieved were disappointing.

Meanwhile, and in spite of the limited investment devoted to it—about 3 per cent of all investment in the oil and gas industries in the Fourth Plan, and 8 per cent in the next quinquennium—consumption of natural gas climbed from 3278 million cubic metres in 1945, to 8981 in 1955, and to 28 085 million cubic metres in 1958. A further 3000 to 4000 million cubic metres of associated gas was vented and flared each year over this period.

The foregoing developments in gas production and utilisation are summarised in *Table 6.1*: both in this table and in the remainder of this chapter statistics of reserves, production and consumption are quoted either in milliards (a thousand million) of cubic metres for ease of handling and to simplify overall comprehension, or where more appropriate/necessary in teracalories.

TABLE 6.1 SOVIET GAS PRODUCTION AND UTILISATION: 1928–58 (MILLIARDS CUBIC METRES)

Year	*Non-ass. NG utilised*	*Ass. NG utilised*	*Ass. NG flared*	*Manufactured gas utilised*	*Total gas utilised*	*NG share of PED %*
1928	neg.	0.3	n.a.	neg.	0.3	n.a.
1932	neg.	1.1	n.a.	n.a.	1.1	1.3
1940	0.4	2.9	n.a.	0.2	3.5	1.8
1945	n.a.	n.a.	n.a.	0.1	3.4	2.0
1950	3.6	2.2	3.8	0.4	6.2	2.3
1955	5.9	3.1	3.5	1.4	10.4	2.5
1958	22.5	5.6	4.3	1.8	29.9	5.3

Notes: Non-ass. NG = non-associated natural gas.
Ass. NG = associated natural gas.
PED = primary energy demand (or consumption).
neg. = negligible, i.e. less than 100 million cubic metres.
n.a. = not available.

Some other developments of passing interest during this period, apart from the beginning of an import/export trade in natural gas which will be discussed later, were the Soviet experiments in the late 1940s of using natural-gas-powered railway locomotives equipped with special tenders to carry the gas. How successful and widespread this became is not recorded. Another application introduced was the use of natural gas for automotive purposes.

Here again progress is not recorded in any detail, but by the early 1950s over 150 road trucks in Kuybyshev were said to be operating successfully on natural gas.

On the domestic market front the number of gas consumers rose quite sharply. For example, in Moscow the number of households connected to the gas grid increased from 68 800 in 1945 to 455 600 in 1955; in Leningrad from 12 500 to 241 800. Gas was used increasingly for space heating. Interestingly, the servicing, repair and replacement of domestic gas equipment was carried out free of charge, an innovation which no doubt would be popular in other countries!

To sum up thus far, although some solid progress was made during the Fourth and Fifth Plans and the USSR scored two 'firsts' with the underground gasification of coal and the manufacture of gas from shale, nevertheless progress in developing both manufactured gas and natural gas was below the planners' expectations. The lack of an adequate natural gas transmission system may well have been one of the main reasons for this as indicated by the relatively large quantities of associated natural gas that could not be utilised and had to be vented and flared.

The Emphasis Changes to Natural Gas

By 1955, the USSR's proven natural gas reserves amounted to nearly 700 milliard cubic metres, a dramatic increase from the levels immediately before and just after the Second World War, i.e. 15 milliard in 1940 and just under 40 milliard cubic metres in 1946. In contradistinction, total production of natural gas in 1955 was only 12.5 milliard of which 9 milliard cubic metres was utilised, the balance being flared (see *Table 6.1*).

Most of this gas had been found by accident as a consequence of exploring for crude oil, rather than of a deliberate policy to find gas. It was about this time, or shortly thereafter, that the Soviet planners decided to concentrate greater effort on utilising this growing natural resource. They recognised that natural gas could play a significant role in the industrialisation of the USSR, and that ultimately it could prove to be a more economic source of energy than either coal or oil.

Accordingly, the Sixth Five-Year Plan, subsequently abandoned, and more especially the Seven-Year Plan (1959–65) envisaged a rapid growth in natural gas output and utilisation, substitution of gas for solid fuels (coal), and a consequential greater effort on exploring for new gas reserves. To what extent, if any, the Soviet planners were influenced in their thinking by the huge success achieved by natural gas in the United States will never be known, but the balance of probability is that the development of natural gas in the North American continent had not passed unnoticed: it would be rather surprising if it had.

Already at this time (i.e. around 1955) about 100 cities and towns and some 200 workers' settlements were being supplied with natural gas. Much of this utilisation was concentrated close to the points of production which were located mainly in European Russia and the Ukraine; long-distance transmission of natural gas had yet to be developed on any worthwhile scale. The goals set in the Seven-Year Plan were aimed at increasing natural gas's share of total primary consumption from over 5 per cent in 1958 to about 21 per cent in 1968, an ambitious target by any standards. In more practical terms it meant increasing the supply of natural gas to about 200 cities and towns by 1960, and to over 500 by 1965, embracing a total population of about 83 million. It meant the complete conversion to natural gas of such cities as Moscow and Kharkov and the supply of gas to some 14 million households in cities, towns, workers' settlements and rural areas. To accomplish this the Seven-Year Plan envisaged, inter alia, the construction of over 20 000 km of pipeline so that natural gas reserves in the Volga area, the north Caucasus, the Ukraine and central Asia could be developed and supplied to a much wider area than heretofore.

Although the expansion of the domestic market was an important objective, high priority was given to developing the industrial market and to using natural gas as a petrochemical feedstock. By 1965, the last year of the Seven-Year Plan, the aim was to have built up a total gas market of some 150 milliard cubic metres of the composition given in *Table 6.2*.

TABLE 6.2 THE TARGET FOR GAS UTILISATION IN 1965

Market sector	*Milliard cubic metres*	*Percentage*
Domestic	14.4	9.6
Chemical feedstock	8.5	5.7
Iron and steel	27.5	18.3
Non-ferrous metals	10.8	7.2
Engineering industries	12.0	8.0
Cement manufacture	8.4	5.6
Power generation	39.8	26.5
Other uses	17.1	11.4
Field and pipeline uses	11.5	7.7
TOTAL	150.0	100.0

In fact actual production (excluding flared gas) realised in 1965 was only 129.4 milliard cubic metres (including 1.7 milliard cubic metres of gas manufactured from coal and shale), some 20 milliard short of the planned target. Nevertheless, compared with utilised production in 1959 of 37.3

milliard cubic metres, this represented an average annual increase (a.a.i.) over the Seven-Year Plan of 19.5 per cent, far higher than the a.a.i. achieved for coal (1.9 per cent) and for oil (9.4 per cent) over the same period. To put these Seven-Year Plan and the two previous plan figures into better perspective, *Table 6.3* shows how natural gas's contribution to total primary energy consumption developed over the period 1951–65.

TABLE 6.3 APPARENT PERCENTAGE CONSUMPTION OF PRIMARY ENERGY (EXCLUDES NET IMPORTS/EXPORTS)

Primary energy	*Fifth Plan 1951*	*1955*	*Sixth Plan* 1956*	*1958*	*Seventh Plan 1959*	*1965*
Coal	64.3	63.1	61.5	56.7	54.2	41.4
Oil	17.5	20.6	22.6	25.3	27.2	34.2
Natural gas	2.3	2.5	2.9	5.3	6.2	14.9
Peat	4.7	4.2	3.5	3.3	3.4	2.2
Shale	0.5	0.7	0.7	0.7	0.7	0.7
Wood	8.6	6.4	6.0	5.2	5.0	3.1
Hydro-power	2.1	2.5	2.8	3.5	3.3	3.5
TOTAL	100.0	100.0	100.0	100.0	100.0	100.0

* Originally planned to cover the period 1956–60 but subsequently abandoned.

As will now have been gathered, the Seven-Year Plan was really the take-off period for the natural gas industry in the USSR. Targets were set high, and although they were not entirely realised in practice, natural gas was on the move in a big way and was one of the fundamental elements of the planners' aspirations to revitalise the Soviet economy. In this respect the role for natural gas was to be reinforced and underlined in subsequent plans.

The Development of Natural Gas Reserves

The bedrock on which the Seven-Year Plan and subsequent plans were based was, as far as natural gas was concerned, on the expectation of large new discoveries. Before discussing the progress achieved in this regard it is first necessary to explain briefly how the USSR classifies its oil and natural gas reserves.

Common Western industry practice is to regard 'proven' reserves as that quantity of natural gas which geological and engineering data demonstrate with reasonable certainty to be recoverable from known oil and gas reservoirs.

'Probable' reserves are those quantities of gas for which there exists a probability of 50 per cent that they will materialise: such reserves are usually allocated to some conjectural part of the field or reservoir as yet undrilled or incompletely evaluated. 'Possible' reserves are those quantities of gas thought to be producible but where the chances of their being realised are considered to be about 25 per cent. Finally, there are 'expectations from future discoveries' which is self-explanatory.

The Soviet system of reserves classification is rather different: they have six main categories. Category A are reserves that have been fully explored by productive wells. Category B embraces gas reserves in areas with favourable prospects borne out by a commercial output from at least two wells. Categories A and B, therefore, correspond roughly, but not precisely, with proven reserves. Category C_1 defines reserves in locations shown to be favourable to the accumulation of gas and where a commercial flow of gas has been obtained from at least one well in the area. Again this can be loosely equated with 'probable' reserves. Category C_2 relates to new structures in known gas-bearing provinces and is approximately similar to 'possible' reserves. Finally, Categories D_1 and D_2 are to all intents and purposes the same as 'expectations from future discoveries'. Having drawn these rough comparisons, it is necessary to state that Western industry practice itself is by no means as uniform as the above might imply.

Published data on Soviet gas reserves usually combine Categories A and B and frequently C_1 as well. For all practical purposes, given that there is no means of verifying the data in question, either of these combinations can be regarded as indicating the order of magnitude of the USSR's proven reserves.

At the beginning of 1959 Soviet natural gas reserves (A + B) stood at just under 1000 milliard cubic metres. The original target over the Seven-Year Plan period was to prove up an additional 3040 milliard cubic metres, but this was subsequently scaled down to 2100 milliard cubic metres. However, even this revised target was not achieved and actual additions to reserves in the A + B categories over the period amounted to 1735 milliard cubic metres. This meant that net of past production, A + B reserves by end 1965 stood at some 2125 milliard cubic metres. But if C_1 reserves are included then the end 1965/early 1966 total was 3565 milliard cubic metres.

It is perhaps appropriate to digress here for a moment. Both as to consumption (utilisation) and additions to proven reserves, the goals set in the Seven-Year Plan were not realised. Many Western commentators tend to stress this, but in all fairness, while the goals turned out to be overly ambitious, full recognition should be given to the actual results that were achieved. Most politicians and planners tend to set their sights too high, a characteristic not confined to any nation or political party, and what in fact the USSR achieved on the natural gas front should not be minimised on this account—a more than doubling of its reserves portfolio and a nearly four-fold increase in natural gas utilisation, figures which speak for themselves.

The growth in reserves ($A + B + C_1$) in the 20-year period since the mid-1950s and their geographic disposition by main region are illustrated in *Table 6.4* which also underlines the importance of the Siberias in more recent years.

TABLE 6.4 NATURAL GAS RESERVES ($A+B+C_1$) BY MAIN REGION

Main region	1955 mrd m^3	%	1965 mrd m^3	%	1975 mrd m^3	%
European USSR, incl. the Urals	642	92	1 803	50	4 200	16
Central Asia and Kazakh SSR	41	6	1 163	33	3 400	13
Eastern Siberia Oblast, Far East and Yakut ASSR	5	1	147	4	2 000	8
Western Siberia Oblast	4	1	455	13	16 200	63
TOTAL USSR	692	100	3 566	100	25 800	100

Note: The figures quoted represent the position at the end of the years stated.

Details of the disposition of reserves in 1976 are not precise, but from such information as has been published it is clear that the major additions to reserves since 1970 have been in western Siberia and central Asia. At the 13th World Gas Conference in 1976, S. A. Orudjev, Minister of the USSR Gas Industry, stated that 'proven' reserves at end 1975 were about 27 000 milliard cubic metres. The significance of this figure is apparent when one realises that it represents more than one-third of the world's total proven natural gas reserves. Subsequent reports suggest that the USSR's gas reserves (Categories $A + B + C_1$) at end 1975 were somewhat lower than this (i.e. 25 800 milliard), but such discrepancies are only to be expected as some of the large fields in western Siberia have yet to be fully appraised, quite apart from the problems of classification already discussed.

On the basis of reserves of 25 800 and a production of 289 milliard cubic metres in 1975, this represented a reserve: annual production ratio of nearly 90:1. The corresponding figures for 1976 were 28 000, 321 milliard and 87:1.

The Super-Giants and Future Expectations

The Soviet Union is blessed uniquely with a number of gas fields to which the media have given the name 'super-giants'. The fields that so far merit this appellation are shown in *Table 6.5*.

Urengoy, where reserves estimates have been progressively raised from 2000 to 4000, to 5000 milliard cubic metres since the field was discovered, is now generally accepted as being the world's largest non-associated natural gas

TABLE 6.5 MAJOR SOVIET GAS FIELDS

Name of field	*Location*	*Date of discovery*	*Year of first production*	*Assessed reserves (milliard m³)*
Urengoy	Tyumen oblast	1966	1978	±5 000
Yamburg	Tyumen oblast	1969	n.y.d.	±3 000
Zapolyarnoye	Tyumen oblast	1965	n.y.d.	±2 000
Orenburg	Orenburg oblast	1967	1971	2 000
Medvezh'ye	Tyumen oblast	1967	1972	1 550
Shatlyk	Turkmen SSR	1968	1973	1 500
Bovanenko	Tyumen oblast	1971	n.y.d.	±1 500
Kharasavei	Tyumen oblast	1974	n.y.d.	±1 000

Notes: n.y.d. = not yet decided.

Except for Orenburg, Medvezh'ye and Shatlyk, which now have several years of production history, the reserves quoted for the above will obviously be subject to verification/amendment in the light of further appraisal drilling. Meantime the figures quoted should only be regarded as orders of magnitude and not as 'proven' reserves.

Medvezh'ye was reported to have reached its peak rate of production (65 milliard cubic metres per annum) in 1977.

Assessed reserves do not, where appropriate, exclude past production.

field. As production only started in April 1978, it will be interesting to see in a few year's time what the ultimate recoverable reserves of this field are then assessed to be.

In addition to these super-giants, the USSR also claims to have several other fields with reserves in excess of 400 milliard cubic metres, and over 30 fields each with reserves in excess of 100 milliard cubic metres—not super-giants but even so large fields in their own right by international standards. What is not in question is that with a few notable exceptions the main bulk of the USSR's gas reserves lies in the prolific Tyumen Oblast. The exploitation of these reserves cannot be divorced from the local conditions and, therefore, they are of no immediate value until such time as the necessary production, treatment, and transmission facilities have been built to enable this gas to be brought to where it can be used effectively, either within the Soviet Union itself, or to convenient locations for export.

Much the same sorts of problems will probably have to be faced in developing new gas resources as most of these are thought to be located close to or within the Arctic Circle. Estimates of the USSR's ultimate resources of natural gas span a wide range from two to four times the size of existing reserves. Such estimates are by definition highly speculative and further discussion of them would be out of keeping with the general theme of this book. Suffice it to say expectations are considerable, and in the opinion of some independent geologists are likely to be substantially larger than ultimate expectations of Middle East natural gas.

Expectations and Achievements in Gas Production 1966–78

This period covers the Eighth (1966–70) and Ninth (1971–75) Five-Year Plans, and the first three years of the Tenth (1976–80) Plan. In the Eighth Plan one of the principal objectives was to mount a major economic development of known oil and natural gas resources, timber, and metal deposits in areas east of the Urals, notably in Siberia. For natural gas this meant not only a substantial exploration, appraisal, and production effort, but also a major construction programme of large-diameter pipelines and compressor stations to bring these remotely located gas resources to the main areas of consumption in the Soviet Union. It was realised that this work programme would carry on in the Ninth Plan, and indeed would spill over into the Tenth Plan as well.

Ambitious goals were set in the Eight and Ninth Plans which in the event were not realised as far as the latter years of each plan were concerned. Although the failure to realise these goals came under heavy criticism, not only by Western commentators, but also within the Soviet Union itself, nevertheless considerable progress was achieved for all that. Undoubtedly the planners had set their sights too high, but whether this was deliberate, or a lack of proper appreciation and practical experience of the time it would take and the problems to be overcome in these remote, inhospitable, and sparsely populated parts of the country, cannot be determined precisely even with hindsight. Certainly no other country in the world had attempted at that time to embark upon a similar programme of this magnitude under such adverse climatic and physical conditions.

As progress fell behind the original goals, revisions were made at various times during the course of the Eighth and Ninth Plans and as might be expected these revisions were then more or less met.

Achievements against the Tenth Plan cannot be monitored in quite the same way as the original plan did not specify yearly production goals, only a range of 400 to 435 milliard cubic metres by 1980. However, following the good performance achieved in 1976, an increase of 32 milliard cubic metres over 1975, yearly targets were set in early 1977 and the upper end of the range for 1980 became the preferred target. Progress in 1977 and 1978 has also been encouraging, in fact since 1974 annual increments have not been less than 25 milliard cubic metres. If this is kept up, the goal for 1980 should be capable of realisation (*Table 6.6*).

But whether the targets now set for 1979 and 1980 will be met or even exceeded will depend more on the rate at which pipelines, compressor stations and gas treatment plants can be completed, than on the availability of

TABLE 6.6 NATURAL GAS PRODUCTION: ACTUALS VERSUS PLANS: 1966–80 (MILLIARDS CUBIC METRES)

Year	*Original plan*	*Plan revisions*	*Actual (non-associated)*	*Actual (associated)*	*Actual total*	*Annual increase*
1966	142	148	125	18	143	15
1967	158	160	138	19	157	14
1968	170	171	149	20	169	12
1969	191	184	159	22	181	12
1970	225–240	198	175	23	198	17
1971	211	211	187	25	212	14
1972	229	—	195	26	221	9
1973	250	238	210	26	236	15
1974	280	257	233	28	261	25
1975	300–320	285	261	28	289	28
1976	318	—	n.a.	n.a.	321	32
1977	not given	342	n.a.	n.a.	346	25
1978	not given	370	n.a.	n.a.	372	26
1979	not given	400				
1980	400–435	435				

Notes: The above, by definition, excludes manufactured gas.
The figures in this table have been rounded deliberately and thus not too much precision should be attached to the accuracy of the last digit for any year.
It is not known to what extent, if any, associated gas which was vented and flared is included or excluded from the above production data. However, it is thought that flared gas is generally excluded.

sufficient gas production capacity. The supply of large-diameter pipe from both Soviet and Western sources will probably be just about adequate, but compressor units and sophisticated gas treatment equipment remain in short supply. The availability and timely construction of these critical items of equipment, and the skilled labour required to install and operate them, are likely to be the main hurdles to be overcome in the coming years if planned targets are to be realised.

Finally, what *Table 6.6* does not show, although it has been touched on elsewhere, is that whereas in 1965 over 86 per cent of the Soviet Union's total natural gas supply of 128 milliard cubic metres came from regions lying to the west of the Urals, notably the Ukraine with 31 per cent, by 1975 the western region's contribution had dropped to under 55 per cent, but, of course, of a much higher total of 289 milliard cubic metres. With many of the older fields in the Ukraine, Stavropol' Krai and Krasnodar' Krai now past their best, the motor for future growth lies firmly and increasingly in the Tyumen Oblast, Turkmeniya, Uzbekistan and some other regions east of the Urals—one of the important exceptions to this being Orenburg.

Transportation

Reserves are one thing, bringing them to where they can be usefully put to work is quite another thing—the more so in the Soviet Union than in most other countries, given the shift of the bulk of natural gas reserves from west to east of the Urals in recent years.

Previously it has been described how most of the country's natural gas was utilised in the early days close to the points of production. In 1946, only some 20 per cent of total utilised gas was transported over any distance of consequence. It was not until 1956 that the first 720 mm (28 inch) line was completed, the largest at that time in Europe. This was a 1254 km line from Stavropol to Moscow. By the beginning of 1960 the total length of main transmission pipelines had grown to some 16 500 km.

The first half of the 1960s saw a rapid expansion of the transmission system consisting mainly of lines of 820 mm (32 inches) or less, although several 1020 mm (40 inch) diameter lines were also brought into service. Inevitably, the main effort at that time was concentrated on linking gas fields located to the west of the Urals with local main centres of consumption and with other cities and towns in central European Russia, the Ukraine, the Baltic Republics and elsewhere: the gasification of Moscow was, of course, a prime objective.

Further expansion of the country's main transmission system took place in the following years, particularly during the Ninth Five-Year Plan. But as with other countries reviewed in this book, it is not proposed to discuss here these developments in any detail as the majority of pipelines built during the 1960s and early 1970s posed no special new problems which had not been encountered in other countries. The situation in the Netherlands was, however, rather different from the norm and merited individual treatment (see Chapter 5), likewise later on in this chapter some details are given of the Bratstvo and Soyuz pipelines because of their importance to the USSR's export trade. Otherwise up to the mid-1970s the progress achieved in developing the Soviet Union's pipeline system is largely self-evident from *Table 6.7*.

What cannot be glossed over quite so readily is the opening up of the west Siberian reserves by pipeline, and the introduction of increasingly larger-diameter pipe sizes. October 1974 saw the commissioning of the first high-capacity line from the northern Tyumen Oblast to the Urals, from there to the Volga area, and finally on to the Moscow region. This and subsequent pipelines from Tyumen presented quite new challenges of a technological and logistical nature as such lines had to traverse vast areas of sparsely populated permafrost country. Laying pipelines above ground, on the ground, or beneath the surface in permafrost country had their various merits and disadvantages in regard to construction techniques and maintenance; safeguarding the environment was another important consideration. These various methods were all tried with varying degrees of success, depending on

TABLE 6.7 DEVELOPMENT OF THE SOVIET UNION'S GAS PIPELINE SYSTEM: 1961–77 (ALL IN KILOMETRES OF PIPE AT END YEAR)

	Transmission pipelines		*Distribution pipelines*		*Total*
Year	*Annual*	*Cumulative*	*Annual*	*Cumulative*	*cumulative*
1961	4 345	25 328	n.a.	17 634	42 962
1962	3 164	28 492	1 469	19 103	47 595
1963	4 297	32 789	6 068	25 171	57 960
1964	4 292	37 081	4 934	30 105	67 186
1965	4 904	41 985	4 374	34 479	76 464
1966	4 376	46 361	4 590	39 069	86 430
1967	6 228	52 589	4 902	43 771	96 360
1968	3 499	56 088	5 756	49 727	105 815
1969	7 071	63 159	6 060	55 787	118 946
1970	4 360	67 519	6 841	62 628	103 147
1971	4 751	72 270	7 332	69 960	142 230
1972	6 414	78 684	8 608	78 568	157 252
1973	4 637	83 321	7 631	86 199	169 520
1974	9 313	92 634	9 367	95 566	187 200
1975	6 068	98 702	6 831	102 397	201 099
1976	4 296	102 998	6 502	108 899	211 897
1977	8 298	111 296	7 869	116 768	228 064

the actual conditions encountered. More recently the preferred method has been to bury such lines. Considerable research and development is going on to effect year-round cooling of the gas to sub-zero temperatures close to the soil temperature to avoid localised heating-up of the permafrost. It is also claimed that gas chilling would have the added advantage of increasing through-put capacity. The first facility to do this is now under construction at Urengoy and will have the capacity to cool some 30 milliard cubic metres per annum down to about minus 10°C.

The Soviet Union was the first country in the world to utilise pipes of 1020, 1220 and 1420 mm (40 to 56 inch) diameter. The use of these pipe sizes rose from 13.8 per cent in 1965 to 39.2 per cent of the total transmission system in 1975. Even larger pipes of 1520 mm (60 inch) and more have been considered, but have yet to be introduced on a practicable scale.

Other innovations or developments on the way include the so-called Sever-1 automatic welding machine, which moves through the pipe itself pulling the next length of pipe into position for welding. This was successfully used on the 56 inch Soyez line and reduced individual joint welding times from about an hour to ten minutes. Trials with multilayer or laminated pipe have been successful and this type of pipe should be introduced on a commercial scale in

6.2 Construction of a large-diameter gas line somewhere in central Asia, USSR, 1978 (*Oil and Gas Journal*)

1980. This pipe has the merits of lower steel content, lower capital and operating cost, greater strength, and improved crack-arresting properties over conventional pipe.

Over the last 15 years or so pipeline construction has gone ahead with commendable zeal and encouraging results in spite of the severe adverse conditions through which many of the more recent larger lines have been laid. Immense logistical difficulties have had to be overcome and skilled labour and equipment have been thinly spread. In these circumstances, achievements against planned goals should perhaps not be examined too critically. Where the achievements have tended to fall short of target is not so much in the laying of pipe per se, but in ensuring that the pipe in question is always of adequate quality. More particularly, if a criticism needs to be levelled at the development of the Soviet Union's transportation system, it should more properly be directed at the inadequate processing/treatment plant available at pipeline input points, and especially at the failure to provide sufficient compressor station capacity in harmony with the completion of new trunk lines in order to bring the latter up to an efficient capacity at an early stage.

Market Sectors

In previous sections of this chapter some broad indications were given of how gas has been utilised in the Soviet Union in earlier years. Since the early 1960s rather more data have become available, but still insufficient to make a detailed analysis of certain market sectors.

As far as the domestic market is concerned, no differentiation is made in published statistics between those customers supplied with natural gas and manufactured gas through the gas distribution grid, and those supplied with liquefied petroleum gas (LPG) in pressurised cylinders. However, statements made by Soviet officials indicate that roughly 45 per cent of total domestic gas consumers are connected to the gas grid, and 55 per cent are supplied with LPG.

Table 6.8, which shows the growth in domestic gas consumers and the types of gas supplied in teracalories, cannot, therefore, be compared directly with the progress achieved in many other countries where the LPG market is usually accorded separate statistical treatment.

If the assumption is made that about 45 per cent of the 48.2 million domestic customers in 1977 were supplied with piped natural gas, this implies an average annual consumption rate per customer of approximately 5000 Mcal. For the same year the rates achieved in the United States were 29 200 Mcal per customer, in the Netherlands 22 600 Mcal, and in the United Kingdom 11 700 Mcal. However, this comparison is distorted by the fact that in the Soviet Union a significant but unspecified proportion of the domestic market only uses natural gas directly for cooking and water heating. Space-heating requirements, which account largely for the high average annual rates of gas consumption in some other countries, are frequently supplied in the Soviet Union by indirect means such as district heating schemes, waste heat from local factories, or electricity from gas-fired power stations. A further complication is that generally speaking the living area that needs to be heated in the average Soviet domestic dwelling is physically a lot smaller than its counterpart in Western Europe or North America.

However, ignoring such disparities, it is evident that considerable progress has been made in bringing gas, in one form or another, to the domestic

TABLE 6.8 DOMESTIC MARKET: CUSTOMERS AND GAS SUPPLIES FOR SELECTED YEARS

Year	*No. of customers (000s)*	*Natural gas (Tcal)*	*Manufactured gas (Tcal)*	*Coke-oven gas (Tcal)*	*Refinery gas (Tcal)*	*LPG (Tcal)*	*Total (Tcal)*
1961	4 306	21 462	800	713	819	2 760	26 554
1965	10 349	42 520	396	161	219	7 564	50 860
1966	12 533	45 996	271	236	155	9 708	56 366
1970	23 376	61 642	315	nil	nil	19 920	81 877
1971	26 752	66 052	319	nil	nil	25 960	92 331
1975	41 722	95 837	363	nil	nil	40 056	136 256
1976	45 255	101 018	339	nil	nil	42 888	144 245
1977	48 226	107 832	335	nil	nil	45 756	153 923

Note: Refinery gas was phased out of the domestic market in 1966 and coke oven gas in 1969.

market. It would appear that over 20 million homes are now connected to the gas distribution grid. Apart from what this implies in the creation of the necessary infrastructure, i.e. distribution and service mains, metering and control equipment, etc., other reports indicate that there has had to be a complementary production of 4 to 5 million cooking appliances and a further 1 million or so water heaters per annum in recent years. The availability of these and other types of appliances may well be a limiting factor for the future growth of the domestic market. In contradistinction, the price of gas for domestic use is nominal by western standards, servicing charges are believed to be nil or very low, and with the cost of connection being borne by the state, collectively these make gas an attractive fuel to use and this must tend to push up demand.

Analysis of other market sectors for gas in the Soviet Union is a vastly more complicated affair and it is only possible to make some general observations. *Table 6.9* shows the breakdown by main market sector for natural gas, excluding other types of gas which are also used to a lesser extent for certain applications in some of these sectors.

From this table it is clear that since the early 1960s the largest single use for natural gas has been power generation representing roughly one-quarter of the total quantity of distributed natural gas. The use of gas for power generation helps to achieve two purposes. First, it reduces atmospheric pollution in and around the main cities; the Soviets claim that this is the main reason why Moscow is today one of the cleanest cities in the industrial world. Second, in those areas where there is no convenient storage facility, power

TABLE 6.9 UTILISATION OF NATURAL GAS BY MAIN MARKET SECTOR: 1961–77 (000s TERACALORIES)

Market sector	*1961*	*1965*	*1970*	*1975*	*1976*	*1977*
Domestic	21	43	62	96	101	108
Commercial	40	98	151	198	237	254
Industry	317	634	843	1 201	1 251	1 322
Chemicals	21	108	86	166	189	221
Iron and steel } Non-ferrous metals }	69	174	233	286	298	308
			19	49	51	52
Engineering	48	121	158	194	208	216
Building materials	108	144	152	192	195	206
Other industry	71	87	195	314	310	319
Power generation	153	336	419	570	637	688
Transport and agriculture	1	2	12	27	36	41
Losses	12	31	37	65	72	61
Total distributed	544	1 144	1 524	2 158	2 334	2 474

stations are supplied with natural gas on an interruptible basis and this helps to balance and improve the load factor of the nation's gas supply system.

For other applications the relative abundance of natural gas has resulted in it becoming the principal source of energy for several industries. For example, gas is used to produce over 85 per cent of open-hearth furnace steel and blast-furnace pig iron, nearly 45 per cent of rolled steel, about 60 per cent of total cement production, and over 80 per cent of the country's requirements of ammonia. None of these figures are precise, but they are indicative of the industrial importance of natural gas.

Having examined the internal situation within the USSR, let us now turn our attention to the natural gas import/export trade with other countries.

Natural Gas Imports

In 1957, Afghanistan signed an exploration agreement with the Soviet Union which resulted in the discovery of a natural gas field at Hodja-Gugerdag, near Shibarghan in northern Afghanistan, in April 1961. Subsequently, further gas deposits were found nearby at Yatim Tag. Agreements were concluded between the two governments in 1963 for the provision of production facilities, the construction of a 32 inch diameter, 98 km pipeline to the Soviet–Afghanistan border at Kelif, and the supply of 58 milliard cubic metres over a 25-year period. Exports to the USSR were scheduled to start at the rate of about 1.5 milliard cubic metres per annum, building up to around 3.5 milliard cubic metres per annum over the contract period.

Part of these gas reserves, equivalent to about 1 milliard cubic metres per annum, was set aside for local power generation and fertiliser manufacture at Mazar-i-Sharif, about 90 km to the east of Shibarghan. The export pipeline to Kelif, and from there some 270 km within the USSR to Mubarek, in Uzbek SSR, to link up with the existing Soviet central Asia and Bukhara–Urals pipeline gas systems, was completed in 1967 and exports of gas to the USSR commenced in that year. Since 1970 exports to the USSR have continued at a fairly constant annual rate of around 2 milliard cubic metres; cumulative supplies over the period 1967 to 1977 inclusive amount to 20 milliard cubic metres.

The USSR's second natural gas import contract was concluded with Iran on 13 January 1966. It concerned the supply of about 140 milliard cubic metres of associated gas from the Marum and Agha Jari oil fields in southern Iran over a 15-year period commencing in 1970. This deal also involved the provision of Soviet industrial and military equipment and a steel mill, with the USSR also being responsible for building the northern Iranian section of the 40 inch diameter export pipeline—from Saveh to Astara on the Soviet–Iran border, a distance of about 500 km—as well as a 30 inch diameter, 112 km spur line from Saveh to Tehran. Construction of the southern Iranian section, comprising a 42 inch diameter pipeline of some 650 km, was undertaken by European and American contractors.

From Astara the USSR built a 220 km pipeline to link up with their existing trans-Caucasus pipeline system which supplies associated natural gas to various centres in Azerbaijan SSR, Armenian SSR, and Georgia SSR, from the declining oil fields near Baku and Groznyy. This system is also connected to gas fields at Maykop, Stavropol and Kanevskaya, and through Rostov ultimately to Moscow. Iranian gas, therefore, served the dual purpose of supplementing declining associated gas supplies in the southern Caucasus and releasing supplies from the more northerly Caucasian fields for transmission to main centres en route to Moscow. For Iran it meant an economic return for gas that would otherwise be flared. Both in logistic and economic terms it was a sensible deal for both countries.

Exports through the Iranian Gas Trunkline, known generally as IGAT-I, started in 1970 and by 1972 had reached an annual rate in excess of 6 milliard cubic metres. Cumulative deliveries over the period 1970–77 inclusive have exceeded 45 milliard cubic metres. In the latter part of 1978 exports of natural gas to the USSR were reduced drastically, and for a time were shut off completely, as a result of the civil disturbances in the Iranian oil fields which cut back oil and associated gas production. As a result some towns in the southern Caucasus suffered severe shortages of gas supplies, until it was possible to reverse the transmission of Soviet gas produced in that region to make good the deficiency in Iranian gas imports. A resumption of Iranian gas supplies at the rates achieved prior to the autumn of 1978 will depend, inter alia, on how quickly Iranian oil production builds up to somewhere around its former level.

Table 6.10 summarises the USSR's imports of natural gas realised over the period 1967 to 1977.

TABLE 6.10 USSR IMPORTS OF NATURAL GAS: 1967–77

Year	*From Afghanistan*		*From Iran*		*Total imports*	
	mrd m^3	Tcal	mrd m^3	Tcal	mrd m^3	Tcal
1967	0.21	2 271	—	—	0.21	2 271
1968	1.50	16 464	—	—	1.50	16 464
1969	2.03	22 266	—	—	2.03	22 266
1970	1.97	21 581	0.73	8 037	2.70	29 618
1971	1.91	20 935	4.27	46 848	6.18	67 783
1972	2.16	23 732	6.22	68 281	8.38	92 013
1973	2.08	22 782	6.59	72 300	8.67	95 082
1974	2.17	23 716	6.91	75 753	9.08	99 469
1975	2.17	23 774	7.26	79 676	9.43	103 450
1976	1.90	20 842	7.04	77 269	8.94	98 111
1977	1.80	19 725	7.13	78 177	8.93	97 902
TOTALS	19.90	218 088	46.15	506 341	66.05	724 429

Note: The above statistics in teracalories (Tcal) are derived from the United Nations' *Annual Bulletins of Gas Statistics for Europe*. Conversion to milliard cubic metres assumes an average gross calorific value of 10 970 kilocalories per cubic metre.

On 30 November 1975, a unique trilateral deal was concluded between the National Iranian Gas Company (the state-owned gas entity), Sojuzgazexport (the Soviet gas export agency), and three European customer companies—Ruhrgas A. G., Gaz de France and Oesterreichische Minearloelverwaltung A.G. of Austria: in December 1976, Czechoslovakia also became a party to this arrangement. Under this deal the USSR will receive some 17 milliard cubic metres per annum for 20 years from 1983 via a second (IGAT-II) 1400 km trans-Iranian pipeline which was under construction and was due for completion in 1981 when initial supplies are scheduled to commence. This Iranian gas was to be consumed within the USSR and in turn some 14.6 milliard cubic metres per annum of Soviet gas was to be supplied to Czechoslovakia, and via Czechoslovakia to Austria, West Germany and France: the 'balance' of 2.4 milliard cubic metres being retained within the USSR for own use. However, because of recent events the prospects of this deal being implemented are now remote.

As already indicated, although the USSR has the world's largest natural gas resource base, these import arrangements with Afghanistan (and Iran) are not only mutually beneficial in economic terms, but aid the logistics of gas supply in a country as vast as the USSR. There are, of course, some political overtones inherent in any deals of this nature, but these were perhaps of secondary consideration in these particular instances where the benefits of mutual co-operation were, originally at least, fairly evenly balanced.

Exports of Soviet Gas to East Europe

The first export of Soviet gas is now generally accepted to have taken place in 1946. This was to Warsaw, Poland, from the Stryji field in that part of the Ukraine which was formerly Polish territory and incorporated into the Soviet Union in 1946. As the pipeline in question existed prior to 1946, this first export of Soviet gas in fact really reflects a territorial change of ownership rather than the commencement of a new grass-roots export business; the latter was not to be realised until 21 years later.

On an equivalent calorific value basis Soviet imports of natural gas and locally produced manufactured gas each currently provide about one-quarter of Poland's total gas supplies, the balance comprises indigenous natural gas and methane gas from coal mines. Although there were 3.4 million domestic gas consumers in Poland at end 1977, average consumption rates are low as gas is not yet used for space heating to any worthwhile extent: the main uses for gas are the chemicals industry (33 per cent of total consumption) and the iron and steel industries (about 30 per cent). By the end 1976, 155 cities and towns out of the 372 connected to the gas grid had been converted to natural gas.

Until the mid-1960s annual imports of Soviet natural gas by Poland did not exceed 350 million cubic metres. Thereafter they rose first to about 1 milliard, and then from the mid-1970s to around 2 milliard cubic metres per annum; they are expected to double again from 1979 onwards.

Czechoslovakia was the second East European country to receive Soviet gas. Imports started in 1967 and have built up progressively to reach an annual rate of nearly 4 milliard cubic metres in 1977. Locally produced natural gas and gas manufactured largely from brown coal amount to a further 1 and 3.5 milliard cubic metres per annum respectively. The main outlets for gas are the chemicals and iron and steel industries; particular importance is accorded to the former, which receives a preferential price of only a quarter of that charged to other industrial consumers. There are approximately 1.9 million domestic consumers accounting for only about 10 per cent of total natural gas sales as they are still supplied in the main with manufactured gas. Conversion of the domestic market to natural gas is going ahead fairly slowly and it is unlikely that more than a quarter of this market will have been converted by 1980.

In 1981, Czechoslovakia should have received its share of Iranian gas under the previously discussed trilateral deal between Iran, the USSR and three West European countries. For Czechoslovakia this involved the supply of 3.6 milliard cubic metres per annum for 20 years as from 1983, after an initial 2-year build-up period. Whether it will now receive Soviet gas instead remains to be seen. In any event conversion of the country's gas system wholly to natural gas will take many years.

The German Democratic Republic received its first supplies of Soviet gas in April 1973. In 1977, supplies of Soviet gas at a little under 3 milliard cubic metres constituted about 40 per cent of East Germany's total gas consumption, the balance comprising roughly one-third locally produced natural gas and rather more than one-quarter manufactured gas. Indigenous natural gas has a very high nitrogen content with a resultant gross calorific value (according to a 1974 report, the most recent available) of only some 4000 kilocalories per cubic metre. Official reports indicate that a substantial proportion of locally produced natural gas is used for power generation because of its high nitrogen content.

There are approximately 3.5 million domestic consumers, but only a very small proportion of them have been converted to natural gas. Most domestic consumers in East Germany are supplied with either manufactured gas or natural gas reformed to town gas quality. Further increases in the production of manufactured gas are planned which would seem to suggest that the domestic market will remain largely based on low calorific value gas for some years to come.

Little can be said about Bulgaria as hardly any information of consequence has been published in recent years by the authorities on the gas market in that country. All that is known of relevance to this account is that Bulgaria has been importing Soviet natural gas since 1974. This gas is supplied from Shebelinka, the largest gas field in the Ukraine. Delivery is affected by a pipeline passing Odessa on the Black Sea to Ismail on the Romanian border, through Romania, to Varna in Bulgaria, and then westwards to Sofia. By 1977, imports had built up to an annual rate of slightly over 2 milliard cubic metres.

Rather more information is available on Hungary which received its first imports of Soviet gas in 1975 to supplement its longer standing but small imports of natural gas (about 200 million cubic metres per annum) from Romania. The domestic market in Hungary is relatively small, barely 800 000 customers, and all but three towns have now been converted to natural gas. Power generation, the iron and steel industry, and the chemicals industry, in that order, are the main users of gas. Hungary currently produces 5 to 6 milliard cubic metres per annum of natural gas, and while local production has risen steadily over the years, the longer-term development of the gas industry is likely to depend to an increasing extent on imports of Soviet gas as local natural gas resources are limited.

Table 6.11 summarises Soviet natural gas exports to East Europe (COMECON countries) over the period 1967–77. The volumetric figures given in this table are derived from the United Nations' *Annual Bulletins of Gas Statistics for Europe* and have been converted from teracalories to milliard cubic metres at an assumed average gross calorific value of 10 970 kilocalories per cubic metre. The latter value has been chosen in order to effect direct comparison with the data previously given (see *Table 6.10*) on gas imported by the USSR. In practice, Soviet natural gas varies in quality and thus somewhat higher volumetric exports of Soviet gas are frequently quoted in literature. However, by using a consistent calorific value for both imports and exports a better 'energy balance' (see *Table 6.13*) can be struck than by relying solely upon published volumetric data comprising gases of varying quality as is usually the case. In this regard the reader would be well advised, when researching any gas data, to exercise caution in drawing too specific assessments based on volumetric quantities where the underlying calorific value of the gas is not stated or known. Unfortunately, in many of the tables given in this book it has not been possible to do this to the extent the author would have wished.

TABLE 6.11 EXPORTS OF SOVIET NATURAL GAS TO EAST EUROPE: 1967–77 (MILLIARD CUBIC METRES)

Year	*Poland*	*Czechoslovakia*	*E. Germany*	*Bulgaria*	*Hungary*	*Total*
1967	1.12	0.17	—	—	—	1.29
1968	1.00	0.59	—	—	—	1.59
1969	0.99	0.89	—	—	—	1.88
1970	0.76	1.08	—	—	—	1.84
1971	1.13	1.25	—	—	—	2.38
1972	1.14	1.47	—	—	—	2.61
1973	1.30	1.80	0.60	—	—	3.70
1974	1.61	2.53	2.20	0.23	—	6.57
1975	1.91	2.81	2.51	0.90	0.46	8.59
1976	1.93	3.50	2.56	1.69	0.76	10.44
1977	2.09	3.86	2.74	2.19	0.76	11.64
TOTALS	14.98	19.95	10.61	5.01	1.98	52.53

In the near future Soviet gas will also be exported to Romania and Yugoslavia. Romania is by far the largest producer of natural gas in Eastern Europe and is one of the 10 largest producers of marketed gas in the world. Production appears to have more or less stabilised in recent years at around 27 to 29 milliard cubic metres per annum, but official statistics post-1975 are not available. It is quite likely that natural gas production in 1977 was substantially below this level as a consequence of the severe earthquake of

March 1977, the epicentre of which was quite close to Romania's main oil and gas fields. How long it took for production to get back to its pre-March, 1977, level is not known.

Apart from a small export commitment of 200 million cubic metres per annum to Hungary for the last 15 years, all Romanian natural gas is consumed internally, of which some 85 per cent is utilised for industrial purposes, mainly for power generation (45 per cent of total consumption) and the chemicals industry. Local production is scheduled to be supplemented by imports of Soviet gas commencing late 1978/early 1979, building up to a rate of 2.5 to 3 milliard cubic metres per annum from the early 1980s.

The export of Soviet gas to the above-mentioned East European countries has considerable economic and logistic advantages for the countries concerned. For the USSR it helps long-term plans for the integration of all European COMECON economies, and for East Europe it represents some degree of diversification of energy supply for economies which hitherto, other than Romania, have relied very largely on coal.

How exports to Poland and Bulgaria were achieved has already been described. For Czechoslovakia, it was the realisation of the so-called 'Bratstvo' or Brotherhood pipeline from the Dashava area in the Ukraine, via Uzhgorod on the Czech–Soviet border, to Bratislava, in southern Czechoslovakia, close to the Austrian border. This export pipeline of some 700 km with a basic diameter over much of its length of 820 mm (32 inches) was completed in August 1967. Subsequently, a northern branch of the Bratstvo pipeline was built from Jablonica in Czechoslovakia to supply East Germany, and a second branch from Uzhgorod to supply Hungary. Meanwhile, on the Soviet side, production from Shebelinka, and later on from Siberia as well, was linked through by pipelines to Uzhgorod to supplement the limited availability from Dashava and Stryji.

Exports of Soviet Gas to West Europe

The completion of the Bratstvo pipeline as far as Bratislava, in a manner of speaking a hop and step from the Austrian border, provided the physical means to develop exports of Soviet gas to Austria. Negotiations between the USSR and Oesterreichische Minearloelverwaltung A.G., the Austrian state oil company, were concluded in January 1968. This contract involved the supply of 30 milliard cubic metres over 20 years and the reciprocal supply by Austria to the USSR of 520 000 tons of 40 inch and 48 inch diameter steel pipe. This was the first supply of large-diameter pipe to the Soviet bloc following the lifting of the North Atlantic Treaty Organisation's embargo on the export of such materials in November 1966. In effect the supply of line pipe, part of which was rolled from Austrian steel plate in West Germany, represented part payment for gas delivered by the USSR. A second gas-for-pipe type of deal was signed on 26 November 1974, increasing the annual quantities of gas to be supplied and extending the supply period to the year 2000.

First deliveries of gas started in September 1968, and by 1976 had reached a rate of about 2 milliard cubic metres per annum. In 1977, Soviet gas imports constituted about one-half of Austria's total gas availability; the balance is mainly locally produced natural gas, plus some small quantities of both manufactured gas and imported natural gas from West Germany which is utilised in the western province of Vorarlberg. It is not known how many of Austria's 900 000 domestic gas consumers have been converted to natural gas: the largest outlet for gas at present is power generation (22 per cent of total gas consumption).

On 1 February 1970, a 20-year contract was concluded between Ruhrgas A.G. and the Soviet Union for the supply at plateau rates of delivery of some 3 milliard cubic metres per annum of Soviet gas to the Federal Republic of Germany. At the same time the reciprocal supply to the USSR of 1.2 million tonnes of Mannesmann steel pipe was agreed. A further 4 milliard cubic metres per annum was contracted for in 1971, and on 29 October 1974, the total annual volume was increased to 9.5 milliard cubic metres.

Exports to West Germany started in 1973 through the Bratstvo pipeline system (the capacity of which had been increased since 1967), via Jablonica to Zlonice in northern Czechoslovakia where extensions to the system bifurcated northwards to East Germany, and westwards to Waidhaus on the Federal Republic border.

West Germany is the largest gas market in Europe with a total consumption roughly 10 per cent greater than the United Kingdom, the next biggest European market. Soviet gas imports comprise less than 10 per cent of total supply availability which is now approaching 50 milliard cubic metres per annum. The largest single component of gas supply is Dutch gas, although locally produced natural gas is substantial and is supplemented by coke-oven gas and other manufactured gases. Soviet gas is, therefore, an important and growing contributor to Germany's supply portfolio, but is likely to remain subordinate in volumetric terms to indigenous and Dutch gas imports for some years to come. Space precludes a more detailed description of this important and fascinating market which, unlike most other gas markets in Europe, is essentially a private industry affair.

Finland began importing Soviet natural gas by pipeline, via Leningrad, Imatra near the Finnish border, to Kouvola and Kotka, under a 20-year contract in 1974. Over 80 per cent of this gas is consumed by the pulp and paper industries located in a relatively small area in eastern Finland. Other industries and district heating systems absorb the balance. The small domestic market of 75 000 consumers is supplied with manufactured gas. To date off-take of Soviet gas has been below contractual expectations, i.e. about

1.2 milliard cubic metres in 1977, because of the depressed state of economic activity in Finland.

Although Italy has substantial natural gas production of its own—some 12 to 13 milliard cubic metres per annum—local production is insufficient to meet Italy's requirements. This has been augmented by imports of LNG from Libya (since 1972), and by Dutch gas via West Germany (since 1974). In May 1974, Italy received its first supplies of Soviet gas via Czechoslovakia (the Bratstvo system) and Austria at Travisio on the Italian–Austrian border. By 1977, Soviet gas imports comprised about 15 per cent of Italy's total availability of approximately 25 milliard cubic metres.

Like most other deals with West European countries the contract with the Soviet Union involved the reciprocal supply of larger-diameter linepipe and related equipment. For Italy this deal represented a further diversification of their natural gas supply portfolio. A more recent 25-year contract with Algeria for the importation by sub-sea pipeline of approximately 12.5 milliard cubic metres per annum, commencing in the second half of 1981, is yet another example of Italy's desire to diversify its supply sources for security reasons.

Unusually Italy has developed a small automotive market for natural gas. The domestic market with 7.2 million customers has now been largely converted to natural gas and represents about 35 per cent of total consumption.

On 6 December 1974, France and the Soviet Union signed a contract for the supply of 78 milliard cubic metres of Soviet gas over a 20-year period, while France undertook to provide the USSR with natural gas desulphurisation equipment and a large aluminium plant. Deliveries of Soviet gas against this contract commenced in 1976, not to France, but to Italy. In exchange France received in 1976 and 1977 the equivalent amount of gas from Italy's contractual supply of Drenthe gas from the Netherlands. This was because the new pipeline system in France to receive Soviet gas at the French–German frontier crossing at Erching–Medelsheim has yet to be completed. This system will be connected to the German transit pipeline and supply Soviet gas to the Paris and Lyons regions and to underground storage at Etrez.

France, like Italy, has an increasing diversity of supply sources which in addition to indigenous natural gas production now comprise Algeria (since 1965), the Netherlands (1967), Norway (1976), and the USSR, with Iran at some future date. At the same time France is active in increasing supply commitments from some of these countries as well as seeking out new sources. The transmission and distribution of gas in France is essentially the responsibility of Gaz de France, a state-owned entity. Conversion of the 8 million domestic customers to natural gas is 90 per cent complete and manufactured gas is being phased out.

Some of the above contractual arrangements discussed above have been modified as to volume and/or duration of supply since they were originally concluded. Also it should be noted that the foregoing statistics and related

TABLE 6.12 EXPORTS OF SOVIET NATURAL GAS TO WEST EUROPE: 1968–77 (MILLIARD CUBIC METRES)

Year	*Austria*	*W. Germany*	*Finland*	*Italy*	*France*	*Total*
1968	0.14	—	—	—	—	0.14
1969	0.78	—	—	—	—	0.78
1970	0.73	—	—	—	—	0.73
1971	1.08	—	—	—	—	1.08
1972	1.24	—	—	—	—	1.24
1973	1.23	0.27	—	—	—	1.50
1974	1.60	1.62	0.34	0.60	—	4.16
1975	1.43	2.43	0.55	1.78	—	6.19
1976	2.11	3.01	0.66	2.83	0.75	9.36
1977	1.89	3.84	0.68	3.82	1.46	11.69
TOTALS	12.23	11.17	2.23	9.03	2.21	36.87

Note: The above data have been derived on the same basis as explained in relation to *Tables 6.10* and *6.11*.

commentary do not reveal that the Soviet Union has entered into a supply commitment with Yugoslavia, due to start in the latter part of 1978, nor the negotiations that have taken place with other prospective customers for Soviet gas such as Belgium and Sweden.

The Orenburg Project

The Orenburg or 'Soyuz' project is probably the most ambitious co-operative task so far undertaken by the COMECON countries and is of particular importance to the Soviet Union's gas export programme. It involves the construction of a 2750 km, 1420 mm (56 inch) diameter pipeline, with 22 compressor stations, from Orenburg, in the Urals, crossing the Volga at Bykovo, to Kremenchug on the Dnieper, and via Bar to Uzhgorod on the Soviet–Czechoslovakian border.

Czechoslovakia, Hungary, Poland, East Germany and Bulgaria agreed in June 1974 to finance and construct approximately one-fifth each of the pipeline and all the compressor stations in their particular sections; the actual pipe and 158 turbine units for the compressor stations being obtained from West Europe. Romania's role in the project was to finance one of the three French-built desulphurisation plants at Orenburg. Orenburg, which was discovered in 1967 and with reserves (Categories $A + B + C_1$) in excess of 2000 milliard cubic metres, is probably the USSR's fourth largest gas field but has a relatively high sulphur content ranging up to 5 per cent. In return for this financing and construction work, the five pipeline construction countries

6.3 Laying a section of the 'Soyuz' 420 mm, 2750 km gas line from Orenburg to Uzhgorod on the Soviet–Czech border in 1978 (*Oil and Gas Journal*)

would each receive 2.8 milliard cubic metres per annum of Orenburg gas, Romania's share being 1.5 milliard, for 12 years commencing 1980. Work began in mid-1975.

By 1977, only Poland and Romania appeared likely to fulfil their obligations. The work commitments of the other four countries were behind schedule because of insufficient labour being dedicated to the task, and in order to maintain the programme Soviet labour had to be drafted in. In compensation for the latter the four countries in question were obliged to step up their financial contributions: it was reported in June 1977 that a consortium of 19 international banks made a loan to COMECON of £293

6.4 A gas-driven compressor, Krummhorn, West Germany, with an output of 780 000 cubic metres of gas per hour at a maximum of 85 bar (1976) (*Ruhrgas A. G., West Germany*)

million largely for this purpose. The pipeline and one compressor station were completed in September 1978, enabling token supplies to be delivered to the Czech–Soviet border. However, the pipeline is not expected to reach its ultimate design capacity of some 28 milliard cubic metres per annum until all the compressor stations are completed which will probably not be realised before 1981.

In the initial years of operation the Orenburg pipeline is expected to be

6.5 A semi-automatic welding process for joining two pipes, first used in Germany by Ruhrgas A.G. in 1973 (*Ruhrgas A. G., West Germany*)

dedicated to supplying export commitments to East Europe—West Europe would continue to be supplied mainly through the existing Bratstvo and related pipeline systems supplemented, in due course, with supplies from the

Orenburg pipeline as this builds up to its ultimate planned capacity. It is estimated that the Soyuz project will have cost the equivalent of nearly US $5000 million when completed.

Liquefied Natural Gas Export Prospects

The possibilities of exporting Soviet natural gas in liquefied form have been under consideration for some years. The two principal prospects are the export of west Siberian gas to the United States east coast and perhaps to northern Europe as well, and the export of east Siberian gas to Japan and the United States west coast.

The former, often referred to as the North Star project, envisages the piping of west Siberian gas (probably from the Urengoy field) some 2500 km to an ice-free port such as Murmansk on the Barents Sea. Annual volumes in excess of 20 milliard cubic metres have been mentioned. However, no concrete progress has yet been made in view of the lack of enthusiasm on the part of the United States Congress for such a venture. Without American participation it is unlikely that European interests would go it alone as West Europe's future requirements of Soviet gas can probably be supplied more efficiently by overland pipeline as is currently the case.

The second LNG project is based on reserves in Yakutsk ASSR, east Siberia, and its possible realisation first depends on discovering a minimum of 1000 milliard cubic metres of which some 825 milliard cubic metres were reported to have been proven at end 1978. Various sites for a liquefaction plant have been proposed. These include Nakhodka and Olga on the Sea of Japan, which are ice-free but involve a much longer pipeline route, and Magadan and Okhotsk on the Sea of Okhotsk, which are much closer to Yakutsk but are not ice-free the year round. Export volumes suggested range from 15 to 30 milliard cubic metres per annum, half each for Japan and the United States. The parties involved comprise Japan Siberia Gas Company, a consortium of various Japanese gas and electricity interests, El Paso Natural Gas and Occidental Petroleum of the United States, and, of course, the Soviet Union. The Japanese have made it clear that they consider it imperative that the venture be developed on a tripartite basis and not as a bilateral deal between Japan and the Soviet Union.

Both these projects pose considerable problems of an economic and logistical nature, and for the Soviet Union the added difficulty of the need to rely upon Western technology, expertise and specialised equipment for liquefaction. Of the two, the Yakutsk project may have the higher probability of realisation because of the limited alternative uses for Yakutsk gas within this region of the Soviet Union, coupled with Japan's desire to develop new sources of gas supply. However, neither of these projects are likely to come to fruition before the late 1980s/early 1990s, and much will in any event depend on how political relationships between the USSR and the USA develop in the coming years.

The Import Export Balance

Table 6.13 shows that, apart from the 4-year period 1970–73, the Soviet Union has been a net exporter of gas since 1946.

TABLE 6.13 SUMMARY OF EXPORTS/IMPORTS OF NATURAL GAS: 1967–77 (MILLIARD CUBIC METRES)

Year	*Exports to E. Europe*	*Exports to W. Europe*	*Total exports*	*Total imports*	*Net (import)/ export balance*
1967	1.29	—	1.29	0.21	1.08
1968	1.59	0.14	1.73	1.50	0.23
1969	1.88	0.78	2.66	2.03	0.63
1970	1.84	0.73	2.57	2.70	(0.13)
1971	2.38	1.08	3.46	6.18	(2.72)
1972	2.61	1.24	3.85	8.38	(4.53)
1973	3.70	1.50	5.20	8.67	(3.47)
1974	6.57	4.16	10.73	9.08	1.65
1975	8.59	6.19	14.78	9.43	5.35
1976	10.44	9.36	19.80	8.94	10.86
1977	11.64	11.69	23.33	8.93	14.40
TOTALS	52.53	36.87	89.40	66.05	23.35

Note: The above excludes exports of Soviet gas to Poland over the period 1946–66 and to that extent the overall export balance since 1946 would be somewhat higher than the 23.35 milliard cubic metres indicated above.

Although the volumes of net exports that have so far been achieved are relatively small, nevertheless such exports are of considerable importance to the Soviet Union. Apart from much needed foreign exchange earnings, almost all the deals with West European countries have involved the reciprocal supply of large-diameter linepipe and/or similar materials and equipment to supplement the USSR's own inadequate manufacturing capacity.

The Soviet Union's performance to date (i.e. up to the end of 1978), as far as gas exports to West European countries are concerned, has been well up to expectations. Although there have been some gas quality problems from time to time, nevertheless contractual obligations have been honoured since exports commenced. Evidence of this satisfactory performance is the apparent willingness of a number of European countries/companies to consider increasing their imports of Soviet gas and/or to conclude new contracts. Obviously there are limits to the extent that any country may wish to become dependent on another country for a major part of its energy supply. However, there would seem to be substantial scope for the USSR to increase its existing gas export commitments before a situation of overdependence is likely to be reached.

Because of the inherent complications in these gas-for-materials type deals with the West it is not possible to assess the true monetary worth to the Soviet Union of its gas export programme—even less so in the case of East Europe. *Table 6.14*, which gives details of the USSR's apparent border prices for gas exports and imports, is, therefore, illustrative rather than definitive and takes no account of the cost of the materials delivered to or by the Soviet Union in full or part exchange for gas supplies.

TABLE 6.14 AVERAGE APPARENT IMPORT/EXPORT BORDER PRICES FOR NATURAL GAS (US ¢ per 1000 cubic feet)

Country	*1970*	*1974*	*1975*	*1976*
Exports to:				
Austria	40	51	119	125
Finland	—	179	184	177
France	—	—	—	96
Italy	—	29	64	52
West Germany	—	53	70	85
Bulgaria	—	57	115	125
Czechoslovakia	46	57	100	133
East Germany	—	56	59	104
Hungary	—	—	117	127
Poland	43	52	110	120
Average for all exports	44	57	91	106
Imports from:				
Afghanistan	17	28	50	47
Iran	n.a.	59	62	58
Average for all imports	n.a.	51	59	56

Note: The data given in the above table are based on official rates of exchange which may not necessarily be the actual rates of exchange applicable in all cases.

Summing-Up

The Soviet Union has made immense strides over the last 15 to 20 years in developing its huge natural gas resources, nevertheless all too frequently

Western observers contrast unfavourably actual Soviet performance against their Five-Year Plan goals. Furthermore, critical comparisons are often made between the gas industries in the United States and the Soviet Union. Many of these comparisons seldom give proper weight to such inherent disparities as the effect of the Second World War on industrial capacity, the state of maturity of the two gas industries in question, differences in terrain, climate and in national economic and political priorities, etc. In the author's opinion such evaluations cannot be wholly objective and relevant unless the starting point on which they are based has a fair measure of common ground; this is hardly the case with the American and Soviet gas industries.

In effect the progress achieved by the Soviet Union has been quite remarkable if all circumstances are taken into account. Undoubtedly, up to the mid-1970s, the Soviet Union consistently set itself unrealistic targets for the gas industry. With hindsight, insufficient account appears to have been taken by the Soviet planners of the vast distances involved, the lack of an adequate infrastructure, and the amount of skilled labour and specialist materials that would be needed to develop those reserves of natural gas which have been discovered in remote, sparsely inhabitated and inhospitable regions of the USSR. Bureaucratic delays, insufficient manpower, inadequate co-ordination, and the attitudes engendered in a rigid state-controlled economy may well have inhibited to some degree the rate of progress achieved. But these and other factors perhaps only serve to underline that comparisons of the Soviet Union with other countries, where the political, social and economic environment is radically different, tend to be counter-productive and frequently misleading. And the fact that since the mid-1970s targets have largely been met or even exceeded, indicates that lessons are being learnt and that more realistic plans are being set.

Perhaps the main criticisms that can be levelled are the time it took for the Soviet planners to appreciate the size and importance of their country's natural gas potential, and the years that were lost in giving priority to developing manufactured gas before this effort was finally abandoned and attention turned to natural gas. The reluctance of the Soviet Union in the past, but this may change, to draw upon the full technical resources and expertise of the western world is understandable, and in some aspects of the business it meant that they had to 're-invent the wheel' with a consequential loss of time and wasted effort.

Ignoring differences of interpretation as to what is meant by proven reserves, there can be little doubt that the USSR has already discovered an abundance of natural gas and has the prospect of finding a lot more. It must surely be only a matter of time before the Soviet Union becomes the world's largest producer, consumer and exporter of natural gas. And in time it is probable that natural gas could become the USSR's largest single source of primary energy supply for non-transportation purposes and its largest earner of foreign exchange. How long it will take for these possibilities to be realised is anybody's guess. Undoubtedly the most crucial factors in any such assessment are the rate at which the Soviet Union can develop its natural gas production capacity, and more particularly expand its gas transmission system in a co-ordinated and harmonious fashion.

In conclusion, it is all too easy to forget that the development of its natural gas resources is but one of the Soviet Union's several priorities, and it will be interesting to see in future Soviet plans the actual priority accorded to gas in relation to other aspirations.

7: Liquefied Natural Gas

Many references have been made in previous chapters to liquefied natural gas, or LNG as it is usually abbreviated. LNG is a relatively new development for the natural gas industry and has a high potential for substantial growth over the next two to three decades at least. The LNG business can be divided into two main categories.

First, the use of LNG for peak-shaving purposes. For this duty surplus pipeline gas is liquefied and stored during periods of normal or low demand, and then regasified and fed back into the pipeline system when demand is high. Apart from providing the gas utility with a much needed supplement to its conventional base-load pipeline supplies during high off-take times, off-peak liquefaction and storage offers the added benefit of improving the load factor for the gas supply system as a whole.

Second, LNG as a base load supply in its own right. LNG is as yet the only economic, technically and commercially proven means of transporting large quantities of natural gas over long sea distances where physical or other constraints preclude the employment of pipelines. This has opened up the prospect of developing natural gas resources in remote locations for export as LNG to markets which have insufficient gas supplies of their own.

Before reviewing how this highly specialised branch of the natural gas industry has developed over the years, it may be helpful first to touch on some of the physical characteristics of natural gas which enable it to be liquefied under certain conditions, also discuss why liquefaction is necessary.

Basic Considerations

Methane, the main constituent of most natural gases as produced in the field, is a gas at atmospheric pressure and temperature. Methane has to be cooled to below minus 82.5°C and compressed to over 45 atmospheres before it will liquefy: no matter how great a pressure is exerted, methane will not remain in a liquid state above its 'critical temperature' of minus 82.5°C, or minus 116°F. For liquefaction at lower pressures, the temperature too has to be reduced. At atmospheric pressure the temperature of liquefaction of methane is about minus 161°C, or minus 258°F.

Other gases usually present in natural gas, in varying but much smaller proportions than methane (e.g. ethane, propane, iso-butane and normal butane), are easier to liquefy. For example, propane can be liquefied at atmospheric pressure by reducing the temperature to about minus 45°C. Alternatively propane can be liquefied at ambient temperatures by the application of relatively modest pressure even in tropical areas where the ambient temperature may be as high as 45°C.

The reason for liquefying natural gas is that liquefaction is currently the only means of reducing natural gas to a sufficiently manageable volume for bulk transport by ship. Liquefied methane occupies approximately 1/600th of the space occupied by gaseous methane at normal temperature and pressure.

Although natural gas can be liquefied by cooling alone, it is more economically liquefied under a combination of part pressure and part cooling. Transport of methane under pressure, however, would necessitate using very heavy pressurised cargo tanks. Because the cost of these pressurised tanks would more than offset the savings in not cooling to minus 161°C, almost invariably under commercial conditions the practice is to reduce the temperature during the liquefaction process to produce LNG at minus 161°C at a pressure at or fractionally above atmospheric pressure.

In a fully refrigerated LNG ship, cargo tanks do not need to be of pressure vessel form so that although spherical and cylindrical tank forms are employed in some designs, the majority of designs are based on rectangular flat-walled tanks tailored to fit and to make optimum use of the available ships' hold spaces. Some of the technical aspects of liquefaction, LNG ship design, etc., will be touched on later in this chapter.

Early Research and Experimentation

The technology for liquefying gases dates back nearly 200 years. M. van Marum and A. P. Troostwijk of the Netherlands were probably the first to liquefy air under laboratory conditions as long ago as 1787, although it was not until 1897 that Karl P. G. von Linde (1842–1934), a German chemist, devised a continuous process of liquefaction and liquefied oxygen, nitrogen and air on a commercial scale. Around 1823, Michael Faraday (1791–1867), the English chemist and physicist, liquefied chlorine, carbon dioxide, sulphuretted hydrogen and sulphur dioxide. Faraday is generally regarded as being the pioneer of the branch of physics that became known as cryogenics.

During the next 50 years or so other scientists variously liquefied hydrogen, oxygen, nitrogen, carbon dioxide, methane and nitric oxide. Some of the more notable researchers at that time were the Englishman, James Prescott Joule (1818–89), and Thomas Andrews (1813–85), an Irish physical chemist, who discovered that for every gas there was a temperature above which pressure alone could not liquefy it: he called this temperature the 'critical point'. Louis Paul Cailletet (1832–1913), a French chemist, liquefied oxygen, nitrogen, and carbon monoxide in 1877, as also did Raoul Pierre Pictet (1846–1929), a Swiss chemist, at around the same time. But it is Cailletet who can probably claim to have been the first to liquefy methane, also in 1877.

The last of the common gases to be liquefied, helium, was liquefied in 1908 by the Dutch physicist, Heike Kamerlingh-Onnes (1853–1926) in the cryogenic laboratory he established at Leiden University. By the beginning of the First World War the basic technology of liquefying gases by a combination of pressure and cooling had been established in a number of academic institutions in several countries.

The Birth of the LNG Industry

Natural gas was first liquefied on a practical scale by the United States Bureau of Mines in 1917. The prime objective of this effort was to separate helium from natural gas in order to obtain helium for use as the lifting agent in airships. Having extracted the helium the liquid methane was regasified immediately and the resultant 'cold' used as a refrigerant to help cool the incoming supplies of natural gas being fed to the plant.

In 1917, Godfrey Cabot of Boston, West Virginia, was granted a patent for an apparatus to liquefy natural gas under a combination of pressure and cooling. Cabot's company, the Liquid Fuel Company, built a small plant at Elizabeth, West Virginia, where natural gas was successfully liquefied. The intention was to use the resultant LNG for welding, among other purposes, but this was not satisfactory and the plant was abandoned in 1921.

Interest in LNG revived in the 1930s when H.C. Cooper, president of the Hope Natural Gas Company of West Virginia, was 'intrigued to the greatest possible extent' by the fact that LNG produced 600 times its own volume when gasified. In 1937 this company commenced intensive studies of liquefaction, storage and regasification, and in 1939 built a pilot plant at Cornwall, West Virginia. This plant utilised a cascade system of refrigeration and liquefied 400 000 cubic feet per day of natural gas. After operating successfully for four months the plant was dismantled in the spring of 1940.

Cleveland—The World's First LNG Peak-Shaving Plant

During the severe winter of 1939/40, the East Ohio Gas Company decided to augment their peak-load capacity. The success of the Cornwall LNG plant resulted in the company deciding to build an LNG peak-shaving plant at Cleveland, Ohio. Work started in September 1940 and the plant commenced operations on 29 January 1941. The cost of the plant came to $1 250 000, including $420 000 for three spherical storage tanks. Liquefaction capacity was 4 million cubic feet per day, storage capacity the equivalent of 150 million cubic feet of gaseous gas, and regasification capacity the equivalent of 3 million cubic feet of LNG per hour. In 1941, total operating costs were about 37 per thousand cubic feet.

The Cleveland plant operated successfully for about 3½ years. In 1943, a fourth storage tank of unusual geometry (toro-segmental cylindrical) and size (90 million cubic feet) was added and brought into service around February 1944. It too operated successfully until 20 October 1944, when it failed. The entire contents of this fourth tank, over 1 million gallons of LNG, spilt into and then over the inadequately bunded area around the tank. Some LNG found its way into the city's storm sewers. The spilt LNG vapourised and caught fire as it came into contact with sources of ignition; sewer systems erupted. One of the three original spherical tanks, its supports weakened by heat, collapsed and added its contents to the conflagration.

The disaster was investigated by the US Bureau of Mines, but the cause of the failure of the fourth tank was never precisely determined. The report suggested that the low-carbon 3.5 % nickel steel used for the inner tank 'might not have been suitable to the particular design of this tank'. No evidence was found that the failure might have been due to improper determination of primary stresses, or to a flaw in the tank, or in the welding of it. Frost spots had been noticed on the outside of the bottom of the tank several months before it failed. Dam walls, not included in the original design, might have restricted ventilation under the tank.

An independent commentator pointed out that whereas the three original tanks were insulated with ground cork, the fourth tank was insulated with rock wool because of a wartime shortage of cork, even though cork was originally specified.

Although this disaster set back the development of the LNG industry in the United States by many years, it at least ensured that major improvements in materials, design, construction, techniques, etc. were introduced—certainly modern LNG plants are vastly different from the Cleveland plant.

The Moscow Plant

In 1947, the closing days of Lease-Lend, a $6 million LNG plant was given to the USSR for erection near Moscow. The plant was designed and fabricated by Dresser Industries, of Dallas, Texas, and shipped to Moscow for assembly by Soviet technicians.

The liquefaction and regasification units were similar to the ill-fated Cleveland plant, but the design of the storage tanks was quite different. These comprised 100 individual stainless steel (chrome-nickel-titanium, 1:18:9) cylindrical vessels, each 10 feet in diameter and 42 feet in length. Supported vertically they were housed in an insulated steel box standing partly below ground level in a concrete pit. Although the manufacturer's design specified 30 inch cork insulation, the Soviets actually used a foam insulation composed of 72.2 per cent formaldehyde, 27.3 per cent urea, and 0.5 per cent sodium acetate. Design capacities were: liquefaction 4.2 million cubic feet per day, storage 160 million cubic feet, and regasification 1.9 million cubic feet per hour.

The plant operated intermittently for several years before it came into continuous operation in 1954. It was originally intended to be used for peak-shaving, but since about 1958 its prime function has been to produce LNG for automotive uses, and to provide LNG for supply by road tanker to up-country satellite regasification plants. Whether this plant is still working, and for what purpose it is now being used, is not known.

The Chicago Union Stock Yards Project

In May 1949, the Chicago District Pipeline Company submitted proposals to the Federal Power Commission (FPC) to build an LNG peak-shaving plant near Joliet, Illinois. Although the FPC issued a certificate of convenience and necessity on 26 July 1949, and stated that the plant was 'suitable, safe and feasible for the purposes proposed, and that the selection of materials, the design of the facilities, and the isolation of the plant were calculated to provide safety in any eventuality reasonably to be expected to occur', the project was abandoned by the company in favour of larger-scale aquifer (underground) storage.

In 1951, a more ambitious project was conceived by William Wood Prince, president of the Union Stock Yard and Transit Company, to ship LNG by barge up the Mississippi from Louisiana to the company's stockyards at Chicago. After considerable research and development, eight barges were planned, each capable of carrying 2500 tons of LNG (or say 120 million cubic feet of gas) in five tanks of about equal capacity. Towed in pairs and taking up to two weeks to make a round trip, the barges would have been capable of delivering about 18 million cubic feet of gas per day.

In the event the project did not materialise, but one LNG transportation barge, the *Methane*, was built to the design of W. L. Morrison of the Lake Forest Research Institution, Illinois. This barge was 264 feet long and 52 feet wide and embodied a unique method of tank insulation utilising 12 inch thick pre-compressed balsa-wood panels. The barge also provided for 'boil off' gas to be used as fuel for the tow-boat's engines.

In 1954, while this work was in progress, Continental Oil Company joined with Union Stock Yard to form the Constock Liquid Methane Corporation (later Constock International Methane Limited) to develop the barge project on a much larger scale with trans-ocean shipments in mind. In February 1956, tests with the liquefaction unit mounted on a barge pumping LNG into two balsa-wood-lined tanks of the transportation barge were carried out at Bayou Long, Louisiana. These tests were observed by the Bureau of Mines and the US Coast Guards and in May 1956 were declared successful. Further tests were held off the Louisiana coast, together with experiments of a prototype barge with liquid nitrogen at minus 320°F. By the spring of 1957, tentative requirements for the waterborne transport of LNG had been drawn up by the US Coast Guards.

Ocean Transport of LNG

The foregoing experiments attracted the attention and interest of the British Gas Council to the point at which in late 1957 it was announced that preparations were underway for the first trial shipments of LNG from the United States to England. This led to a joint effort between the British Gas Council and Constock Liquid Methane Corporation to convert the 5000 ton dry-cargo freighter *Normati*, renamed *Methane Pioneer*, to transport approximately 2200 tons of LNG in five tanks insulated with laminated balsa-wood. Plans for the conversion of this ship were prepared by the marine architects, J. J. Henry Company of New York.

Meantime, liquefaction and LNG loading facilities were built by Constock on the Calcasieu River, 8 miles south-west of Lake Charles, Louisiana, to be fed with natural gas from nearby fields owned by Continental Oil Company, while the British Gas Council built reception facilities and LNG storage at Canvey Island, Essex, on the Thames Estuary.

After sea trials and tests in late 1958, *Methane Pioneer* left Lake Charles on 28 January 1959 with a full cargo for Canvey Island, delivery of which was completed on 20 February. Over the next 14 months, six further cargoes were transported through all sorts of weather and successfully delivered at Canvey Island. The safe shipment of natural gas in liquid form had been clearly demonstrated.

The Algeria–UK/France LNG Projects

Early in 1960, the Royal Dutch/Shell Group of Companies acquired a 40 per cent interest in Constock, which was then renamed Conch International Methane Limited. The balance of shares in Conch were held 40 per cent by Continental Oil and 20 per cent by Union Stock Yards. The first task of Conch

7.1 *Aristotle*, formerly *Methane Pioneer*, which carried the world's first trans-ocean LNG cargo in January 1959

was to open negotiations with Société d'Exploitation des Hydrocarbures d'Hassi R'Mel (SEHR), the then French owners of the huge Hassi R'Mel gas field in Algeria. These negotiations resulted in the formation of Compagnie Algérienne du Méthane Liquide (CAMEL), owned jointly by Conch and French interests, to build a liquefaction plant at Arzew. Later, the Algerian government acquired a 20 per cent interest in CAMEL, leaving Conch and the French interests with a 40 per cent stake each.

On 3 November 1961, the British Minister of Power, Richard Wood, approved the Gas Council's request to import LNG from Algeria for an initial period of 15 years. The gas purchase contract between CAMEL and SEHR was signed on 12 December 1961. And on 14 September 1962, the foundation stone of the Arzew liquefaction plant was laid by Ben Bella, the Algerian Prime Minister.

The Arzew plant was designed to liquefy somewhat in excess of 150 million cubic feet of natural gas per day in three separate units, or 'trains' as they are usually called. Of this 100 million cubic feet per day was for the UK contract, and the balance of 50 million cubic feet per day for processing by CAMEL for Gaz de France for shipment to Le Havre. The three-train Arzew liquefaction plant (including processing capacity for France) was based on a cascade cycle of refrigeration, constructed by Technip & Pritchard, and was reported to have cost about £31 million; at that time it represented one of the largest single investments ever made on the African continent.

Responsibility for buying LNG f.o.b. Arzew, shipping and reselling it to the Gas Council c.i.f. Canvey Island was vested in British Methane Limited, owned 50 per cent by Conch and 50 per cent by the Gas Council. For this project two ocean-going LNG carriers were built to Conch design in consultation with J. J. Henry Company. *Methane Princess*, built by Vickers-Armstrong at Barrow, Lancashire, and owned by Conch Methane Tankers Limited, was launched on 22 June 1963. *Methane Progress*, built at Belfast, Northern Ireland, by Harland & Wolff, and owned by an associate of Houlder Brothers, was launched on 19 September 1963. Both ships have a capacity of some 12 000 tons (or say 27 400 cubic metres) of LNG carried in nine cargo tanks insulated with balsa-wood and glass-fibre. Because the specific gravity of LNG is only about half that of crude oil, these ships compare in size to a conventional oil tanker of around 28 000 tons. They were reported to have cost some £4 750 000 each, about twice that for an oil tanker of comparable size. For the Arzew to Le Havre complementary movement a slightly smaller ship, *Jules Verne* with a capacity of 25 800 cubic metres, was built to the Worms self-supporting tank design by Chantier de la Seine and owned by Gaz Marine.

While the gas production facilities at Hassi R'Mel, the 24 inch diameter, 250 mile pipeline to Arzew, the liquefaction plant and ships were being built, the Gas Council were extending their LNG storage capacity, regasification and other facilities at Canvey Island at a cost of £3.5 million as those provided for

7.2 *Methane Progress* built by Harland & Wolff Ltd, Belfast, in 1963. Capacity 27 400 cubic metres. Owned by Methane Tanker Finance Ltd, and operated by Shell Tankers (UK) Ltd. Continuously employed since 1964 on the Algeria–UK project (*Shell*)

the original imports by *Methane Pioneer* were inadequate for this much larger long-term import project. At the same time a methane pipeline system extending some 250 miles from Canvey Island to Leeds, with branch lines to serve eight of the area gas boards en route, was constructed at a cost of £10 million. Similarly, Gaz de France put in hand the building of a new LNG reception and regasification terminal at Le Havre, together with the necessary pipeline linkage to their existing gas system.

On 26 September 1964, *Methane Princess* loaded at the CAMEL liquefaction plant, and on 12 October 1964 she delivered the world's first commercial LNG cargo to Canvey Island. Not long after this *Methane Progress* entered into service and in February 1965 the first cargo by *Jules Verne* was delivered to Le Havre.

Between October 1964 and December 1977 over 1100 cargoes of LNG were delivered to Canvey Island and Le Havre. During this time the project, both as a whole and in all its separate but inter-linked phases, performed extremely well. Apart from some temporary difficulties between Algeria and France in 1973, from a technical/operational point of view there has hardly been any interruption to supply of major significance over this period. By any standards this project—the world's first international base-load LNG venture—has been a noteworthy success.

Over the years there have been a number of corporate changes. For example, in the early 1960s the Algerian government nationalised the Hassi R'Mel field, and subsequently the various French interests in CAMEL sold their shareholdings to the Algerians. And on 10 May, 1977, Conch's remaining 40 per cent interest in the CAMEL liquefaction plant was acquired by Société Nationale pour la Recherche, la Production, le Transport, la Transformation et la Commercialisation des Hydrocarbures—more simply known as Sonatrach—the Algerian state oil and gas entity. Sonatrach are thus now responsible for all aspects of supply up to f.o.b. Arzew where the LNG is purchased by British Methane and Gaz de France. At the customer end, the Gas Council is now, of course, British Gas Corporation.

LNG Peak-Shaving Makes a Comeback

The Cleveland disaster and the decision not to proceed with the planned plant at Joliet, Illinois, proved to be but temporary set-backs. Whether the success of the *Methane Pioneer* trial shipments and related work was the actual turning point, or but a coincidence of timing, is not known. However, whatever the cause or reason, by the late 1950s/early 1960s serious interest was reawakened in the United States in the prospects of peak-shaving with LNG.

In 1965, three LNG peak-shaving plants were commissioned by Alabama Gas Corporation, San Diego Gas and Electric Company and Wisconsin Natural Gas Company; a fourth was commissioned by Transcontinental Gas Pipe Line Corporation in 1966. The combined capacity of these plants was about 14 million cubic feet of gas per day for liquefaction, some 2000 million cubic feet of storage, and over 300 million cubic feet per day for regasification. In all four cases above-ground storage tanks were installed. In two instances aluminium alloy was used for the inner tank, and for the other two 9 per cent nickel steel. As far as it is known, all these plants have given satisfactory performance.

By the autumn of 1978, the total number of LNG peak-shaving plants (liquefaction, storage and regasification) in the United States had grown to 55. These were supplemented by over 50 so-called satellite plants which have LNG storage and regasification facilities, but as the name implies, have no liquefaction equipment and are supplied with LNG mainly by road tank trucks from the aforesaid peak-shaving plants. In the winter of 1977/78, LNG peak-shaving and satellite plants represented a combined supply deliverability of 7600 million cubic feet of gas per day.

LNG peak-shaving plants and satellite facilities also exist in Canada, Britain, West Germany, France, Spain, the Netherlands and Belgium, with one under construction in Australia. With two exceptions, all these non-American plants have been built since 1970; all have operated without incident.

The Development of International LNG Projects

The second international LNG base-load project to be developed was the Alaska to Japan venture. This was conceived in the early 1960s when there was no apparent shortage of natural gas supplies in the United States. It is a small scheme by present-day standards involving the liquefaction of about 150 million cubic feet of gas per day at Port Nikiski, on the Cook Inlet in southern Alaska. The liquefaction plant has only one train and produces approximately 1 million tons of LNG per annum. The project was promoted by Phillips Petroleum and Marathon Oil and is serviced by two 71 650 cubic metre capacity LNG carriers built by Kockums of Sweden to the Gaz Transport membrane tank design.

First shipment of Alaskan LNG to the sole customers, Tokyo Electric Power Company Inc. and Tokyo Gas Company Limited, was delivered to Negishi, Yokohama, on 4 November 1969. The contract is of 15 years' duration and is due to expire, subject to any extension that might be agreed by 1982, in 1984. The delivered Negishi price for this contract was originally US ¢52 per million Btu, but over the years the price has been revised several times to reflect inflation, changes in energy values, increased operating costs, etc., to over $2 per million Btu by 1978. The project has performed well since its inception with no supply interruptions of consequence.

Next in chronological order was Esso's project to liquefy Libyan gas for shipment to Italy and Spain. From the two-train liquefaction plant at Marsa el Brega, Libya, over 230 million cubic feet of high calorific value gas per day is exported to Italy, and over 100 million cubic feet per day to Spain. The contracts are of 20 years' and 15 years' duration respectively. Four LNG carriers each of approximately 41 000 cubic metres capacity service the project. Although the two-train liquefaction plant at Marsa el Brega was started up in late 1968, because of various technical problems it was not until April 1970 that the first LNG was transferred to storage and only nominal shipments were made in that year. Because of various technical and other difficulties full capacity was not achieved until 1973. This was relatively short-lived and supplies were interrupted during the winter of 1974/75 when problems arose in connection with price re-negotiations as between the customers, the project promoters, and the Libyan government. These were resolved in due course and in more recent years the project has performed at or close to its design capacity.

The Brunei–Japan LNG Project

The Brunei to Japan LNG project represented another important step forward in the LNG international business. This was a much larger project than its predecessors. Initially, it involved a four-train liquefaction plant at

Lumut, Brunei, six LNG carriers each of some 75 000 cubic metres capacity, three reception terminals in Japan, and the supply of about 3.7 million tons of LNG per annum (or say 510 million cubic feet of gas per day) over 20 years. But before the project came on-stream, the supply of an additional 1.5 million tons of LNG per annum was agreed with the customers making a total of 5.2 million tons per annum (720 million cubic feet per day) over 20 years, after an initial 4-year build-up period. In order to supply these supplementary quantities the plant was expanded to a five-train complex, and a seventh LNG carrier was ordered.

The project was promoted by various companies of the Royal Dutch/Shell Group, Mitsubishi Shoji Kaisha (subsequently renamed Mitsubishi Corporation), and the Brunei government, with the former being responsible for technical leadership of the project. The three Japanese customer companies are Tokyo Electric, Tokyo Gas and the Osaka Gas Company Limited.

The corporate structure of the project prior to 1975 comprised:

—Brunei Shell Petroleum Company Limited, a wholly owned Royal/Dutch Shell company, responsible for gas production and the sale of gas at the liquefaction plant gate;

—Brunei LNG Limited, owned 45 per cent each by Shell and Mitsubishi and 10 per cent by the Brunei government, concerned with the liquefaction of gas at Lumut and the sale of LNG f.o.b. Brunei; and,

7.3 One of three 60 000 cubic metre capacity LNG storage tanks under construction at Lumut, Brunei, in 1971 (*Shell*)

7.4 A general view of one of five liquefaction trains of Brunei LNG Ltd's liquefaction plant at Lumut, Brunei (*Shell*)

—Coldgas Trading Limited, owned equally by Shell and Mitsubishi, responsible for the sale of LNG to Japanese customers and for chartering the necessary shipping capacity which is owned and operated by Shell Tankers (UK) Limited.

As a result of various negotiations with the Brunei government the corporate structure since 1977 has been as follows:

—Brunei Shell Petroleum Company Limited, now owned equally by Shell and the Brunei government;
—Brunei LNG Limited, now owned as to one-third each by Shell, Mitsubishi, and the Brunei government; and,

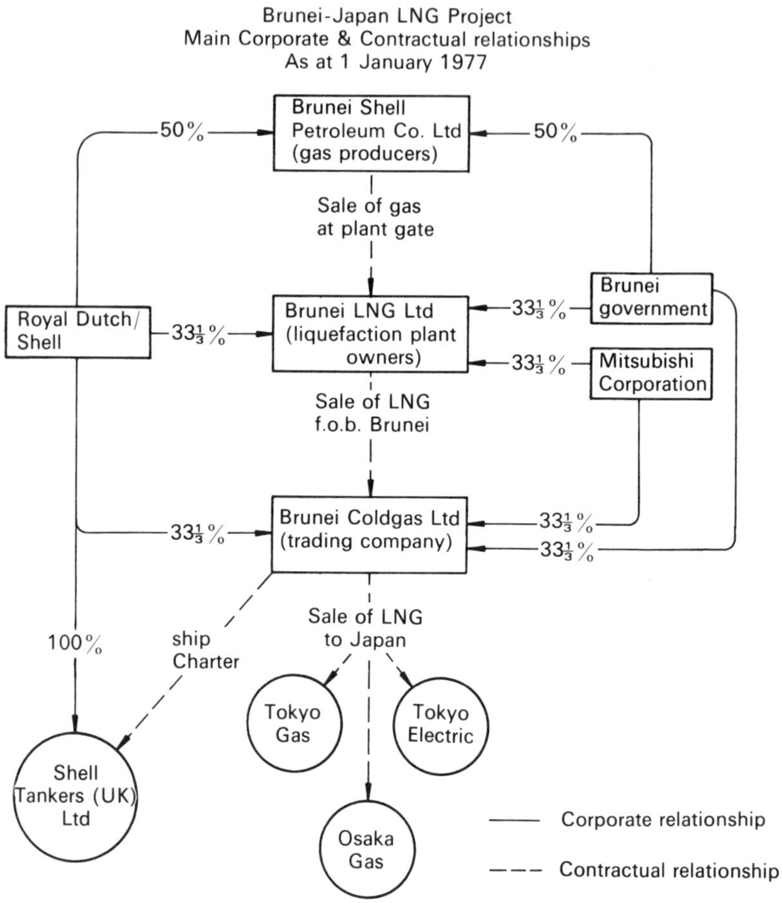

7.5 Brunei–Japan LNG project. Main corporate and contractual relationships as at 1 January 1977

—Brunei Coldgas Limited, which replaced Coldgas Trading, and is owned as to one-third each by Shell, Mitsubishi and the Brunei government.

The Brunei project is currently the world's largest international base-load LNG venture in full operation, a position, however, it will relinquish to the so-called El Paso I project (from Arzew, Algeria to Cove Point, Maryland, and Savannah, Georgia), when the latter becomes fully operational in late 1980 or early 1981. The Brunei project is noteworthy in several respects. Apart from its sheer size at the time it was conceived, it involves a diversity of customers and delivery points which necessitated the development of highly sophisticated computer programmes for ship scheduling, monitoring operational performance, and for a variety of other purposes. One special feature of the project is over-the stern loading of ships at the end of a jetty projecting 4.5 kilometres into the South China Sea at the end of which is a 42 metre high loading crane of unique design.

Reverting to the sequence of main events and activities, Letters of Intent between the project promoters and the Japanese customers were signed in September 1969, whereupon critical equipment for the Brunei plant was pre-ordered; full sales contracts were concluded on 2 June 1970. The construction contract for the plant was awarded to Procon Incorporated and Japan Gasoline Company Limited in January 1970 and construction work at Lumut began with site clearance in March 1970; the plant produced its first LNG in October 1972. The first LNG carrier, *Gadinia*, commenced loading at Lumut

7.6 One of five cryogenics heat exchangers being erected at Brunei LNG Ltd's liquefaction plant at Lumut in 1971. It weighs 100 tons and is 27 metres high (*Shell*)

7.7 LNG carrier SS *Gastrana*, 75 000 cubic metre capacity, loading at Lumut, Brunei, for Japan. The loading crane is approximately 140 feet above sea level and 4.5 km offshore (*Shell*)

on 7 December 1972 and delivered its cargo to Osaka Gas's terminal at Senboku on 15 December—exactly as forecast some two years previously. By July 1974, all five trains of the Lumut plant had produced LNG; by November of the following year all seven French-built ships—five of which have Technigaz membrane (304L stainless steel)-type tanks and two Gaz Transport membrane (36 per cent nickel steel)-type tanks—were in service.

To date the Brunei project has performed extremely well. It commenced operation on schedule, all customer contractual obligations have been met, and no ship has been delayed awaiting product availability.

Other Operational Base-Load Projects

Since 1972, several other base-load LNG projects have been brought into service, some of which are still in the build-up phase and have yet to reach their full design capacity. They are:

—A second project for Gaz de France from Skikda, Algeria, to Fos-sur-Mer in southern France. This is a 25-year contract for the supply of 35 milliard thermies (or 3.7 milliard cubic metres of gas) per annum at plateau rates of delivery. The contract started in 1973, but some pre-contract cargoes were delivered in 1972.

—Algeria's contract with Empresa Nacional del Gas (Enagas), the Spanish state gas entity, for the supply over 23 years of 45 milliard thermies (or 4.8 milliard cubic metres of gas) per annum. Initial supplies commenced in 1976 from spare capacity at Sonatrach's plants at Arzew and Skikda, but full contract volumes are not scheduled to be supplied until 1980.

—The first Indonesian project from Badak, East Kalimantan, to Japan, delivered its first cargo to Osaka on 14 August 1977. And the second project from Indonesia, based on Arun, North Sumatra, delivered its first cargo to Chita on 12 October 1978. These two projects, when fully operational, will supply a total of some 7.5 million tons of LNG per annum (or say 10.5 milliard cubic metres of gas) to five Japanese customers—Kansai, Chubu and Kyushu Electric Power Companies, Osaka Gas, and Nippon Steel Corporation—collectively known as Japan Indonesia LNG Company Limited or JILCO. Pertamina, the Indonesian state entity, and JILCO are involved in both projects with the Huffington group of companies in the case of Badak (Pt. Badak Natural Gas Liquefaction Company), and Mobil Oil Indonesia in the case of Arun (Pt. Arun Natural Gas Liquefaction Company).

—The Abu Dhabi to Japan project involves the supply of about 2.2 million tons of LNG (say 3 milliard cubic metres of gas) and 850 000 tons of LPG per annum from Das Island to the sole Japanese customer, Tokyo Electric, for 20 years. The project promoters, Abu Dhabi Liquefaction Company Limited—comprising the Abu Dhabi National Oil Company, BP, CFP, Mitsui (a Japanese trading house) and Bridgestone Liquefied Gas Company—delivered their first cargo to Tokyo on 14 May 1977.

—The small Algeria–Distrigas of Boston project has in fact been operating on an intermittent basis since November 1971 but the main contract for 20 years for the supply of 12 milliard thermies (or 1.3 milliard cubic metres of gas) per annum did not start until 1978.

The last of the current operational schemes is the previously mentioned El Paso I project based on a major six-train liquefaction plant yet to be completed at Arzew, Algeria. Although the contract between Sonatrach and El Paso Algeria Corporation was signed on 9 October 1969, prolonged delays in obtaining the necessary regulatory approvals in the United States, and the appointment of a new contractor in January 1976, part-way through plant construction, resulted in the first shipment being deferred until March 1978. At full capacity the project will involve the supply of 103.5 milliard thermies (say 10.9 milliard cubic metres) per annum for 25 years. The project will be serviced by nine LNG carriers each of 125 000 cubic metres capacity, three of which are French-built and six in two American shipyards. Three different ship containment systems have been selected—Gaz Transport membrane, Technigaz membrane, and Conch self-supporting.

By 1980, when all the above-mentioned schemes should be at or close to their plateau rates of delivery, international trade in LNG should amount to some 48 milliard cubic metres of gas per annum, which is roughly equivalent to 4800 million cubic feet per day. A summary of the projects comprising this trade is given in *Table 7.1*..

TABLE 7.1 INTERNATIONAL TRADE IN LNG BY 1980

Source	*Market*	*Supplier (f.o.b.)*	*Importer (c.i.f.)*	*Start-up*	*Annual volumes (f.o.b. at plateau) (milliard m³)*
Algeria	UK	Sonatrach	British Gas	1964	1.1 (1)
Algeria	France	Sonatrach	Gaz de France	1965	0.6
Alaska	Japan	Phillips and Marathon	Tokyo Gas and Tokyo Electric	1969	1.6
Libya	Italy	Esso	SNAM (2)	1970	2.4
Libya	Spain	Esso	Enagas	1970	1.1
Algeria	USA	Sonatrach	Distrigas	1971	1.3
Brunei	Japan	Brunei LNG	Tokyo Electric, Tokyo Gas and Osaka Gas	1972	7.2
Algeria	France	Sonatrach	Gaz de France	1973	3.7
Algeria	Spain	Sonatrach	Enagas	1976	4.8
Abu Dhabi	Japan	ADGLC	Tokyo Electric	1977	3.0
Indonesia	Japan	Badak NGLC	JILCO	1977	4.0
Algeria	USA	Sonatrach	El Paso	1978	10.9
Indonesia	Japan	Arun NGLC	JILCO	1978	6.5
				TOTAL	48.2

Notes: It is assumed that the Algeria–UK contract will be extended when the original 15-year period expires at end 1979.
SNAM is a subsidiary of ENI, an Italian state entity, and is responsible for the importation and transmission of almost all natural gas sold in Italy.

Apart from the start-up of a third scheme for the United States, the so-called Trunkline project, scheduled for October 1980, and possibly some expansion of exports from Indonesia (ex Badak), no other projects are scheduled to come on stream by 1980. Actual quantities supplied f.o.b. in 1980 may indeed be somewhat lower than 48 milliard cubic metres, if there are any slippages in the build-up of existing projects. In any event delivered quantities will be lower than those indicated in *Table 7.1* as in practice all the major containment systems of LNG carriers are designed for average boil-offs in the loaded condition in the range of 0.2 to 0.3 per cent of cargo capacity per day.

Although this book is concerned essentially with historical facts, it may nevertheless be appropriate to conclude this chapter with brief comments on pricing, how the LNG business may develop, and on some of the technical and other considerations inherent in the LNG international business.

Prices and Pricing

Unlike the international oil business, where 'posted' or 'marker' prices tend to set the basis around which actual prices for particular crude oils and products are established, there are no similar pricing systems of general application for LNG. Many factors, such as competitive energy values, the pricing of indigenous pipeline gas supplies, very much higher transport costs for LNG than for oil (which for LNG also vary sharply with the distance involved), different fiscal regimes and many other considerations, militate against the establishment of relatively uniform world-wide prices for LNG.

Obviously no LNG project will be developed unless it is expected to produce a satisfactory financial return for both the promoters of the project and for the host government concerned. For private industry, this return will need to be commensurate with the political, commercial and technical risks involved, and adequate to attract and service all the financing required, including third-party loans. For host governments, the return will probably comprise a package of income including tax on those phases of the venture under the government's jurisdiction, income as a commercial partner in the venture, royalty and so on.

From the customer's point of view, his ability to pay for LNG will be conditioned by his market mix both as to types of customers and (firm) supplies; load factor and quality considerations; the cost and availability of such alternative supply possibilities as may be open to him; the extent to which 'roll-in' pricing of higher-cost LNG imports with lower-cost pipeline gas is possible and/or permitted; etc.

An illustration was given above (for the Alaska to Japan project) of how and why one particular LNG c.i.f. price has moved over the last decade. Whilst it is possible to deduce average yearly c.i.f. prices for various other LNG projects from official statistics, it is not customary for project promoters to publish actual prices as these are usually confidential between the contractual parties concerned. The main exception to this generality is in the case of projects directed at the United States where full contractual details, including prices, have to be filed with the appropriate regulatory bodies and as such become matters of public record. Even in such cases full exposure of the ex-field and into-liquefaction plant transfer prices may not be necessary if the filing was, for instance, based on an f.o.b. or c.i.f. sale.

One country that has endeavoured to introduce a measure of price uniformity for LNG export contracts concluded since the mid-1970s is

Algeria. Here the approach has been to adopt an f.o.b. floor and/or base price of US ¢ 130 per million Btu as at 1 July 1975. This base or floor price is subject to periodic reviews, is indexed usually partly or wholly with competing oil prices, and in a number of instances is adjusted to reflect currency changes. Actual details vary from contract to contract, but the general pricing philosophy underlying these contracts is much the same. However, Algerian contracts concluded before 1975 were made on a discrete basis and do not conform to this pattern, although changes are being made to some of these older contracts as and when circumstances permit to bring them into line with the pattern established since 1975.

The Organisation for Petroleum Exporting Countries (OPEC) has given a lot of thought to possible pricing systems for LNG exports, but the practical difficulties of applying any universal system to a business of this nature where costs, market values, shipping distances, and other factors can vary to such a marked degree, have so far proved insoluble.

In general, LNG prices, like indigenous gas or for that matter international pipeline gas prices, tend to be concluded on an individual basis appropriate to the supply source and market concerned. With specific supply sources/projects dedicated to specific customers/markets, this discrete approach to pricing is both logical and necessary as should be apparent from other aspects of the LNG business described elsewhere in this chapter.

LNG Systems

An LNG project usually has at least four distinct phases of activity:

(1) gas production, initial treatment and transmission by pipeline to the liquefaction plant;
(2) liquefaction, including further gas treatment, storage and ship loading;
(3) shipping and trading, delivering the LNG to reception terminals; and,
(4) unloading of LNG, storage, regasification and send-out of gas to customers by pipeline.

Each phase of activity may be corporately separate, but in practice an LNG project is one integrated or closed-loop system stretching from the gas well to the ultimate consumer. There is a dynamic interaction between each phase of the project, and if any one element in the chain is not ready in time or fails for any reason, the viability of the whole project may be in jeopardy; at the very least the economics of these complex, capital-intensive projects will suffer adversely until such time as all the links in the chain are operating on a consistent and harmonious basis. In a sense an LNG project is akin to a pipeline gas venture, if one regards the shipping phase as a 'flexible pipeline'.

Any LNG project to be economic must also operate at a high load factor, at least as far as the production, liquefaction and shipping phases are concerned. However, at the receiving end some elasticity can be achieved if storage capacity over and above that necessary to accept LNG shipments on a regular basis is provided. In this way LNG can be stored and regasified at a variable rate. The economics of regasifying LNG imports on a regular as opposed to a variable basis will depend to a large extent on the cost of LNG storage versus the costs (and availability) of alternative supplies in the market concerned; there are no golden rules, each case must be examined on its merits. However, as already stated, from the suppliers' point of view, LNG shipments must be regular and continuous for operational and cost-effective reasons.

Unlike the international oil business, LNG does not lend itself to spot cargo or flexible trading. Each LNG project is specifically tailored and optimised to supply a given annual quantity to a given outlet or outlets over a 15 to 25-year contract period. The economics of this high-cost business preclude the degree of flexibility of supply that the oil business enjoys: the lack of surplus liquefaction capacity and the limited availability of reception terminals world-wide, all of which are dedicated to base-load supply schemes, are further constraints to the development of spot-cargo business other than on a very occasional basis.

Production

The production phase of an LNG project is basically no different from a pipeline gas venture, except that possible variations in gas quality over the life of an LNG contract are perhaps more critical, and, of course, adequate reserves and production capacity must exist to keep the liquefaction plant fully loaded at all times. In practice many other considerations of a highly complex commercial, operational, and technological nature have to be taken into account before a gas producer can embark upon a long-term supply commitment for an LNG project, whereas for a pipeline gas project such considerations may not always be so demanding.

Liquefaction

The liquefaction plant is the heart of an LNG project. While various processes are used, the basic principle is the same as that of a domestic refrigerator, although vastly more complicated and sophisticated. Each process train will normally involve purification of the incoming gas feed, dehydration, compression, refrigeration, and liquefaction. Purification and dehydration are necessary to remove undesirable entrained constituents such as carbon dioxide and water. Fractionation provides refrigerant make-up and the separation of any heavy hydrocarbons that may be present that would otherwise solidify at liquefaction temperatures.

There are four elements in the basic cooling cycle: a compressor, a condenser, a pressure-reduction valve and an evaporator. The cycle begins with the compression of a gaseous refrigerant such as propane. As the pressure of the refrigerant increases so does its temperature and its tendency to liquefy. When the warm compressed gas passes into the cold condenser it liquefies, giving off heat as it does, which is in turn carried away by cooling water. The high pressure of the liquefied refrigerant is then released in the pressure-reduction valve. As a result the temperature of the refrigerant drops, taking heat from the incoming warm natural gas which is thereby cooled, and the refrigerant evaporates. As it is not possible to liquefy natural gas in a single propane cycle, it is passed through a series of evaporators each containing a refrigerant of a lower evaporating temperature than the one before.

These cycles form an interlocking system. As well as cooling the natural gas, the evaporator of the first cycle is used to cool the refrigerant of the second, the second of the third, and so on, until the progressively cooled natural gas liquefies. As this arrangement is very complex, many plants utilise mixed-refrigerant cycles of nitrogen and various hydrocarbon refrigerants which span a much wider temperature range than a single refrigerant, thus reducing the number of cycles required.

As LNG is produced it is run to insulated double-walled storage tanks until sufficient product is available to load a ship. This is done through insulated loading lines designed on a fail-safe principle.

The above is a very simplistic description of what, all said and done, is a very sophisticated plant, requiring considerable expertise in its design, construction, and operation, and costing for a modern-day major liquefaction complex upwards of $2000 million. In this regard no mention has been made of the huge power and cooling water requirements, harbour works, and such like, which are an integral part of an LNG plant located more often than not in an area where no infrastructure (roads, housing, schools, hospitals) exists and has to be provided from scratch.

Shipping

If the liquefaction plant is the heart of the project, LNG ships probably attract the most glamour and interest. Since the early *Methane Pioneer* days, the capacity of LNG carriers has grown to the point at which the most common size today is in the range of 120 000 to 130 000 cubic metres. At the end of 1978, 30 LNG carriers with capacities of more than 25 000 cubic metres were in regular service and a further 15 completed, either awaiting project start-up, or yet to be committed to any particular project. Part of this current surplus in shipping capacity arises from the fact that certain independent shipowners failed to appreciate the integrated nature of an LNG project and ordered LNG carriers on a speculative basis without any immediate prospect of firm employment. At $100 to $150 million per ship, depending on capacity, design, where and when ordered, etc., this has proved to be a costly experience for some owners.

A wide variety of LNG ship designs exists or is in various stages of development. However, in terms of designs actually built or on order, the list narrows down to two main groups—self-supporting tanks and membrane tanks—and to six variations thereof:

Self-supporting tanks—Conch prismatic (aluminium); Worms (9 per cent nickel steel); Esso (aluminium); Moss-Rosenberg spherical (aluminium or 9 per cent nickel steel).

Membrane tanks—Gas Transport (36 per cent nickel alloy); Technigaz (304 L stainless steel).

Each system has its merits, but the two market leaders by far are the Gaz Transport membrane and Moss-Rosenberg spherical containment systems with 33 and 22 ships respectively out of a total of 76 ships that have been built or were on order as at end 1978. Including firm orders, France with 31 and the United States with 18 ships, lead the field. Other countries with LNG shipbuilding experience are Sweden, Norway, the UK, Japan, West Germany, Spain, and Belgium. France owes its prominent position to the fact that both the Technigaz and Gaz Transport designs were developed there, also to her foresight in equipping its shipyards at an early stage to build these specialised carriers.

As mild steel becomes brittle and fractures at low stress levels at cryogenic temperatures, LNG containment systems are designed to protect the mild steel of the ship's hull from the outflow of cold from the cargo. Similarly, as the cargo is at its boiling point any heat in-flow from the outside will cause the LNG to evaporate or boil off. The insulation system, therefore, serves the dual purpose of protecting the hull and reducing boil-off to acceptable limits. The same principles apply, of course, to land-based LNG storage tanks. However, LNG ship containment systems bear little resemblance to land tankage: for the former, tanks are tailored to make optimum use of the ships' hold spaces and due account has to be taken in the design of the dynamic effects and torsional flexing of the ship while at sea.

Self-supporting containment systems utilise heavy, thick-walled, structurally independent, metallic tanks. Such systems require substantial quantities of cryogenic materials and are costly to fabricate. But insulation for the vertical walls and tank roof can be made of relatively low-cost material. Conversely, membrane systems rely on a less expensive light, thin, metallic membrane. This is offset by more expensive and sophisticated insulation of a rigid load-bearing nature.

The metallic container, whether of the self-supporting or membrane type, is known as the primary barrier. Depending on the design, a secondary barrier may be necessary to contain temporarily any leakage of LNG through the primary barrier. Generally, all membrane systems require a secondary barrier

7.8 Aluminium LNG cargo tanks to the Conch design under construction at Mobile, Alabama, in 1978. Each tank is approximately 120 feet square, 90 feet high and weighs 850 tons. When completed each tank is lifted in one piece into the ship. There are five such tanks in a 125 000 cubic metre capacity LNG carrier (*Conch International Methane Limited*)

over the whole tank area. For rectangular self-supporting tanks a partial secondary barrier over the lower region of the tank is usually provided, but not in the case of spherical tanks, for which stress problems are not so critical; the secondary barrier for spherical tanks may be limited to a drip tray and a spray shield.

All containment designs allow for the fact that there will be some boil-off of the LNG cargo during the laden voyage. Typically this may be about 0.25

per cent of cargo capacity per day. In most cases this boil-off gas is used as a fuel for ship's propulsion engines, thereby reducing the quantity of bunker oil needed. Some LNG is left in the ship's tanks after discharge, so that the tanks are kept cold until the ship loads its next cargo. This is known as 'cargo heel' and must be sufficient to allow for boil-off on the ballast voyage (generally about 0.15 per cent of cargo capacity per day), plus a safety margin in case the ship is delayed.

In the 14 years of commercial LNG operations, more than 3500 laden voyages have been completed by the world's LNG carrier fleet and no accident has occurred in which significant quantities of LNG have been spilt on to the sea. Nevertheless, research continues to investigate and minimise still further the possible effects of accidental spillages.

Terminals

The function of an LNG terminal is to receive, store, and regasify LNG. A terminal may cost upwards of $500 million depending on the amount of storage provided and land prices. Storage tanks are of the same designs as those in liquefaction plants, but greater numbers may be required to cope with varying patterns in off-take or gas send-out. The heat required for regasification is frequently taken from circulating sea water, supplemented by gas-fired heaters in the winter. LNG reception terminals are far less complex than liquefaction plants, but as they are often located close to main centres of population—simply because that is where the resultant gas is needed—problems of an environmental and similar nature can arise. This is not because the storing and handling of LNG is necessarily any more hazardous or environmentally objectionable than for other hydrocarbons, but it does seemingly attract the attention of environmentalists in spite of its very good safety record to date.

Before concluding this particular section, it should be stressed that the descriptions given above of the main phases of an LNG project are of the briefest possible nature. Each merits a book to itself and many of the technological, commerical, political, economic, fiscal, and contractual complexities inherent in such projects have either been glossed over or omitted. However, it is hoped the reader will have gained a broad impression of what is involved in this relatively new and growing branch of the world's gas business.

What of the Future?

It is an inescapable fact that, on the one hand, the main markets for natural gas, the United States, Western Europe and Japan—the USSR is the exception—are already experiencing or can expect gas supply shortages to develop in the coming years, while on the other hand there are abundant known resources of natural gas in the world at large. LNG provides the only technically and commercially proven means as yet by which some of this remotely located gas can be brought to the markets which will need it.

Many new international base-load LNG projects are in various stages of construction, negotiation, or planning. New supply sources where natural gas is available in quantities greater than the local market can absorb in the foreseeable future are known to exist in various countries, notably in Africa, the Middle East, Central and South America, Asia, Australia, the Sino–Soviet bloc, and in some other specific locations.

It is not possible to predict with any certainty how quickly and how much of these resources will be developed as LNG projects over the next couple of decades. Undoubtedly some projects now being contemplated will not materialise for one reason or another, while some prospects as yet unidentified will emerge. All one can say with certainty is that LNG trade will grow several-fold for several decades to come. Many forecasts have been made—including several by the author of this book—of the rate of growth that might be realised over certain specified periods of time. Such forecasts tend to be rather subjective and conditioned inevitably by the circumstances pertaining when the forecast was compiled. As such they are perhaps inappropriate to quote in an historical account of this nature. However, rather than leave the reader entirely in the air, in the author's opinion it would be surprising if the level of international LNG trade had not grown from its expected level by 1980 of rather less than 50 milliard cubic metres per annum, to something in excess of 150 milliard cubic metres per annum by the year 2000—it could well be substantially greater if some of the historical delays in implementing new projects were minimised, and particularly if the world as a whole begins to experience an increasing shortage in alternative energy supplies in the years ahead as many forecasters have predicted.

Glossary of Some Common Gas Industry Terms

ASSOCIATED NATURAL GAS Natural gas originating from underground structures producing both liquid and gaseous hydrocarbons. The gas may be dissolved in crude oil (SOLUTION GAS), or in contact with gas-saturated crude oil (GAS CAP GAS). In such structures, gas production rates will depend on oil output, with oil usually representing the major part in terms of energy equivalents.

BLUE WATER GAS Also called 'water gas'. Made in a cyclic process in which an incandescent bed of coke or coal is alternately subjected to blasts of air and steam. The gas consists mainly of equal proportions of carbon monoxide and hydrogen and has a gross heat content of about 300 Btu per cubic foot.

BOTTLED GAS Liquefied petroleum gas contained under pressure in cylinders, or 'bottles' as they are sometimes called, for convenience of handling and delivery to customers.

BRITISH THERMAL UNIT Usually abbreviated to Btu. Basically the amount of heat required to raise the temperature of 1 pound of water through 1° Fahrenheit.

BUTANE A gaseous hydrocarbon usually present in small quantities in most natural gases with a gross calorific value of about 3215 Btu per cubic foot. Can be liquefied by the application of modest pressure and/or by cooling.

CALORIFIC (HEATING) VALUE The amount of heat released by the complete combustion of a unit quantity of fuel under specified conditions. For gases, this value may be expressed in Btu per cubic foot, kilocalorie per cubic metre, joule per cubic metre, etc. Practice varies from one country to another. Values may be quoted on a gross or net basis. Briefly, 'gross' means that the water produced during combustion has been condensed to liquid and has released its latent heat, while 'net' signifies that such water stays as vapour.

CARBURETTED WATER GAS Also called 'carburetted blue gas'. Gas resulting from the enrichment of blue water gas during its manufacture by a simultaneous process of light distillate, gas oil or fuel oil gasification. The gas has a gross heat content of about 500 to 550 Btu per cubic foot.

CITY GATE A measuring station, which may also include pressure regulation, at which a distributing gas utility receives gas from a pipeline company or from the main gas transmission system.

COAL GAS or COKE-OVEN GAS A manufactured gas made by destructive distillation (carbonisation) of bituminous coal in a gas retort or as a by-product from a coke oven. Its principal components are methane (20 to 30 per cent) and hydrogen (about 50 per cent). This gas generally has a gross heating value of 500 to 550 Btu per cubic foot. When the process takes place in a closed oven it is generally designated 'coke-oven gas', and when produced in retorts it is called 'coal gas'.

COMBINATION UTILITY A company which supplies both gas and some other utility-type service, e.g. electricity, water, steam, etc. Quite a common type of utility in the United States, but less common elsewhere.

COMMERCIAL CUSTOMER OR MARKET Consumers such as schools, hospitals, hotels, restaurants, offices, shops, etc., which can neither be classified as domestic nor as industrial consumers.

COMMODITY CHARGE That portion of the price, tariff or rate charged to gas customers, based on the total volume of gas consumed or billed.

CONNECTION CHARGE An amount paid by the gas customer in a lump sum, or in instalments, for connecting the customer's facilities to the supplier's facilities.

CONTRACT DEMAND (TAKE-OR-PAY) The volume of gas that the supplier agrees to deliver and, in general, the amount that the customer agrees to take or pay for, hence the expression 'take-or-pay'.

CUBIC FOOT (GAS) and CUBIC METRE (GAS) Common units of measurement of gas by volume. Actual conditions of measurement, e.g. temperature, pressure, water vapour content, etc., vary from one country to another.

CUSTOMER CHARGE A fixed amount to be paid periodically by the customer without regard to the demand or commodity consumption.

DAILY AVERAGE SEND-OUT The total volume of gas delivered for a period of time divided by the number of days in that period.

DAILY PEAK The maximum volume of gas delivered in any one day during a given period, usually a calendar year.

DEMAND CHARGE That portion of the price, tariff or rate charged to gas customers based on the customer's demand characteristics, usually expressed as a fixed sum for a specified period.

DISTRIBUTION GAS UTILITY or PUBLIC DISTRIBUTION COMPANY A company or utility which obtains the major portion of its gas revenue from the operation of a retail gas distribution system. Normally it would not own/operate any major gas transmission pipeline system. Depending on the country concerned, its main retail activities will normally be confined to supplying domestic, commercial and small-volume industrial customers within a designated service area.

DISTRIBUTION SYSTEM The necessary pipeline system, services and equipment which carry or control the supply of gas for a distribution gas utility from the point or points of gas supply (usually the city gate station) to and including the consumers' meters.

DOMESTIC or RESIDENTIAL CUSTOMER or MARKET Consumers which use gas primarily for household or domestic purposes in or about their dwellings.

ETHANE A gaseous hydrogen invariably present in small quantities in most natural gases with a gross calorific value of about 1730 Btu per cubic foot. Can be liquefied by cooling. Apart from the United States, and a few isolated cases elsewhere to date, ethane is almost invariably not extracted from natural gas and as such is a constituent of natural gas (predominantly methane) as distributed to customers.

GAS FIELD A geological structure, formation or reservoir from which non-associated natural gas is produced.

GAS WORKS A plant where a combustible gas (a secondary energy) is manufactured usually from coal (a primary energy). See also MANUFACTURED GAS.

INDUSTRIAL CUSTOMER or MARKET Consumers which use gas primarily for industrial-type applications; in many countries this class of market or customer embraces gas used as a feedstock for the manufacture of chemicals and/or gas used for power generation purposes.

INTERRUPTIBLE GAS Gas made available under agreements which permit the curtailment or cessation of delivery by the supplier at specified times, ambient temperatures, etc.

LINE PACK A method of peak-shaving by withdrawing gas from a section of a pipeline system in excess of the input into that section, i.e. normally the difference between the actual volume of gas in the pipeline at low flow rates (increased pressure) and that at normal flow rates.

LIQUEFIED NATURAL GAS (LNG) Natural gas that has been liquefied normally by cooling to approximately minus 258°F (minus 161°C) at atmospheric pressure.

LIQUEFIED PETROLEUM GAS (LPG) Any hydrocarbon mixture in either the liquid or gaseous state, the chief components of which consist of propane, propylene, normal-butane, iso-butane, butylene or mixtures thereof in any ratio.

LOAD CURVE A graph in which the send-out of a gas system, or segment of a system, is plotted against intervals of time.

LOAD DURATION CURVE A curve of loads, plotted in descending order of magnitude, against time intervals for a specificed period. The co-ordinates may be absolute quantities or percentages.

LOAD FACTOR The ratio of the average load over a designated period to the peak load occurring in that period. Usually expressed as a percentage.

MANUFACTURED GAS Combustible gases derived from primary energy sources by processes involving chemical reaction. For instance, gas produced from coal, coke or oil products.

METER (GAS) A mechanical device for automatically measuring quantities of gas.
METHANE The principal hydrocarbon constituent of most natural gases with a gross calorific value of about 995 Btu per cubic foot.
MILLIARD A thousand million, or 10 to the power of 9; synonomous with the American billion.
MINIMUM CHARGE or BILL A clause in a gas contract which provides that the charge for a prescribed period shall not be less than the specified amount.
NATURAL GAS See associated natural gas and non-associated natural gas.
NATURAL GAS LIQUIDS (NGL) Those hydrocarbons which can be extracted in liquid form from natural gas. Invariably the term NGL embraces propane and all heavier hydrocarbon fractions, i.e. butane, pentane, etc., where present. In some instances it may be taken to mean the inclusion of ethane as well.
NATURAL GAS RESERVES The quantities of natural gas which geological and engineering data demonstrate with reasonable certainty to be recoverable from a known gas or oil reservoir. They represent strictly technical judgements, and are not knowingly influenced by attitudes of conservation or optimism. However, the actual methods of evaluating and quantifying reserves do vary somewhat in practice from one country/company to another.
NON-ASSOCIATED (or UNASSOCIATED) NATURAL GAS Natural gas originating from underground structures from which only gas is produced.
PEAK or PEAK LOAD The maximum load of gas consumed in a stated period of time.
PEAK-SHAVING The practice of augmenting the normal supply of gas during peak or emergency periods from another source where gas may either have been stored during periods of low demand, or manufactured specifically to meet the peak demand.
PRODUCER A company or entity concerned primarily or exclusively with the production of natural gas. In some countries producers may have complementary interests in the transmission of gas or indeed in other related aspects of the gas business.
PRODUCER GAS A gas manufactured by burning coal or coke with a regulated deficiency of air, normally saturated with steam. The principal combustible component is carbon monoxide (about 30 per cent) and the gross heat content is between 120 and 160 Btu per cubic foot.
PROPANE A gaseous hydrocarbon usually present in small quantities in most natural gases with a gross calorific value of about 2480 Btu per cubic foot. Can be liquefied by the application of modest pressure and/or by cooling.
PUBLIC DISTRIBUTION (COMPANY) See DISTRIBUTION GAS UTILITY.
REFORMING The process of thermal or catalytic cracking of natural gas, liquefied petroleum gas (LPG), natural gas liquids (NGL), refinery gas or various oil products, resulting in the production of a gas having a different chemical composition.
REFINERY GAS A gas resulting from oil refinery operations consisting mainly

of hydrogen, methane, ethylene, propylene and the butylenes. Other gases such as nitrogen and carbon dioxide may be present. The composition can be highly variable and the heat content can range from 1000 to 2000 Btu per cubic foot.

RESERVOIR (GAS) An underground rock strata in which natural gas has accumulated and is trapped.

SEND-OUT The quantity of gas delivered by a plant or a system during a specified period of time.

SUMMER VALLEY The decrease which occurs in summer months (in appropriate temperature zones) in the volume of the daily load of a gas distribution system.

SYNTHETIC or SUBSTITUTE NATURAL GAS (SNG) A gas made from coal or oil products which has a calorific value and burning characteristic compatible with natural gas.

THERM 100 000 Btu.

TOWN GAS A generic expression usually taken to mean gas piped to consumers from a gas plant. The gas can comprise both manufactured gas and natural gas used for enrichment.

TRANSMISSION COMPANY A company which operates a natural gas transmission system for the transport of gas in large volumes. Such companies do not normally sell gas direct to end-consumers (except perhaps to very large consumers) but there are exceptions to this generality.

TRANSMISSION SYSTEM The pipeline system of a transmission company or companies.

TRUNK LINE A large-diameter, high-pressure natural gas pipeline for transporting large volumes of gas over relatively long distances.

WATER GAS See BLUE WATER GAS.

WOBBE INDEX The gross calorific value of the gas divided by the square root of the density of the gas compared with air.

Note: The above terms should be regarded as being generally descriptive rather than precise definitions. In some countries some of these terms are precisely determined for local legal or regulatory reasons. Likewise, certain units of measurement included above will be subject to local regulations which impose specific standards and conditions of measurement. It should be noted that considerable efforts are now being made by various international bodies to standardise on terminology and measurements. However, as these have yet to receive world-wide acceptance and as this book is concerned essentially with the past, the above selection attempts to cover only those expressions, etc., which have hitherto been generally used.

Symbols, Abbreviations and Natural Gas Equivalents

Symbols and Abbreviations

The following do not necessarily conform with the International System of Units (SI) as recommended by the International Gas Union. They are nevertheless still widely used by the gas industry as a convenient, if inconsistent, shorthand for everyday purposes.

Btu = British thermal unit
MMBtu = million British thermal units
ft^3 = cubic foot
scf = standard cubic foot
Mcf = thousand cubic feet
MMcf = million cubic feet
Tcf = trillion (10^{12}) cubic feet
Nm^3 = normal cubic metre
mrd m^3 = milliard (10^9) cubic metres

Terminology and Constituents of Natural Gas

	Constituents		
Natural gas ex well	Methane		
	Ethane		NGL
	Propane	LPG	NGL
	Butanes	LPG	NGL
	Pentanes and heavier fractions also referred to as:		NGL
	C_5 +		NGL
	Pentanes plus		NGL
	Natural gasoline		NGL
	Condensate		NGL
	Non-hydrocarbons e.g. water, carbon dioxide, etc.		

LNG = liquefied natural gas
LPG = liquefied petroleum gas
NGL = natural gas liquids
SNG = synthetic (or substitute) natural gas

Natural Gas Equivalents

The tables contain quick-reference equivalents and other factors of general relevance to the natural gas industry. All figures are to be taken as *approximate values* only for use when a high degree of precision is not required.

The approximations in these tables are based upon:

1 for natural gas:
1000 Btu/ft^3 = 9500 kcal/m^3
(Groningen gas 8400 kcal/m^3)

2 for LPG:
an assumed 50/50 propane/butane mixture with (r) or (p) indicating that the LPG is either refrigerated or pressurised.

3 calorific values, MMBtu (gross):

per tonne —LNG 51.8; LPG 47.3;
oil 42.3; coal 27.3
per barrel —LNG 3.8; LPG (r) 4.45;
LPG (p) 4.1; oil 5.8
per cubic metre —LNG 23.8; LPG (r) 28;
LPG (p) 25.8

NATURAL GAS: CUBIC METRE EQUIVALENTS

1 mrd m^3 natural gas per year =	0.04 Tcf gas (38 trillion Btu) 890 000 tonnes oil 800 000 tonnes LPG 725 000 tonnes LNG 1.4 million tonnes coal	per year
	100 million ft^3 gas 17 800 barrels oil 23 200 barrels LPG (r) 25 200 barrels LPG (p) 27 200 barrels LNG	per day

1 million m³ natural gas per day =	0.014 Tcf gas (14 trillion Btu)	per year
	325 000 tonnes oil	
	290 000 tonnes LPG	
	265 000 tonnes LNG	
	500 000 tonnes coal	
	37 million ft³ gas	per day
	6 500 barrels oil	
	8 500 barrels LPG (r)	
	9 200 barrels LPG (p)	
	9 900 barrels LNG	

1 m³ Groningen gas = 0.88 m³ (9 500 kcal)
1 m³ (9 500 kcal) = 1.13 m³ Groningen gas

NATURAL GAS: CUBIC FOOT EQUIVALENTS

1 Tcf natural gas per year =	27 mrd m³ gas (30 mrd Groningen)	per year
	24 million tonnes oil	
	37 million tonnes coal	
	2 700 million ft³ gas	per day
	470 000 barrels oil	
100 MMcf natural gas per day =	0.04 Tcf (37 trillion Btu)	per year
	1 mrd m³ gas (1.1 mrd Groningen)	
	860 000 tonnes oil	
	770 000 tonnes LPG	
	700 000 tonnes LNG	
	1.35 million tonnes coal	
	2.7 million m³ gas (3 million Groningen)	per day
	17 250 barrels oil	
	22 500 barrels LPG (r)	
	24 400 barrels LPG (p)	
	26 300 barrels LNG	

LNG: VOLUMETRIC EQUIVALENTS

1 million tonnes LNG per year =	77 million ft^3 (liquid)	per year
	2.2 million m^3 (liquid)	
	14 million barrels (liquid)	
	0.05 Tcf (gas)	
	1.4 mrd m^3 (gas)	
	1.1 million tonnes LPG	
	1.2 million tonnes oil	
	52 trillion Btu	
	1.9 million tonnes coal	
	140 million ft^3 (gas)	per day
	4 million m^3 (gas)	
	37 500 barrels LNG	
	31 900 barrels LPG (r)	
	34 600 barrels LPG (p)	
	24 500 barrels oil	
1 million m^3 LNG per year =	460 000 tonnes LNG	per year
	6.3 million barrels LNG	
	0.2 Tcf (gas)	
	0.6 mrd m^3 (gas)	
	500 000 tonnes LPG	
	560 000 tonnes oil	
	24 trillion Btu	
	870 000 tonnes coal	
	65 million ft^3 (gas)	per day
	14 700 barrels LPG (r)	
	15 900 barrels LPG (p)	
	17 200 barrels LNG	
	11 200 barrels oil	

m^3 = kilolitre

OIL AND COAL EQUIVALENTS

1 million tonnes oil per year =	1.1 mrd m^3 gas (1.3 mrd Groningen) 1.5 million tonnes coal 815 000 tonnes LNG 890 000 tonnes LPG 0.04 Tcf gas (42 trillion Btu)	per year
	115 million ft^3 gas 3 million m^3 gas 30 500 barrels LNG 26 000 barrels LPG (r) 28 300 barrels LPG (p) 20 000 barrels oil	per day
1 million tonnes coal per year =	0.7 mrd m^3 gas (0.8 mrd Groningen) 640 000 tonnes oil 525 000 tonnes LNG 580-000 tonnes LPG 0.03 Tcf gas (27 trillion Btu)	per year
	75 million ft^3 gas 2 million m^3 gas 19 700 barrels LNG 16 800 barrels LPG (r) 18 200 barrels LPG (p) 12 900 barrels oil	per day
10 000 barrels oil per day =	0.6 mrd m^3 gas 500 000 tonnes oil 780 000 tonnes coal 0.02 Tcf gas (21 trillion Btu)	per year
	58 million ft^3 gas 1.5 million m^3 gas	per day

(The tables have been derived and adapted from Shell International Gas Ltd's *Natural Gas Equivalents*, 1978)

Selected References

The following are some of the principal references consulted by the author in compiling this book. Additional information was obtained, or consulted for verification purposes, from a wide variety of publications, journals, and newspapers too numerous to list in entirety.

Where apparent inconsistencies and discrepancies existed, the author selected what he considered to be the most reliable/accurate source of information. For ease of reading and comprehension, no attempt has been made to litter the text with a multiplicity of references.

The United Kingdom

Allcock, J. F., *Natural Gas Purchasing*, The Institution of Gas Engineers, 1978.

British Gas Corporation, *Annual Report and Accounts* (various years).

Chandler, Dean and Lacey, A. Douglas, *The Rise of the Gas Industry in Britain*, British Gas Council, 1949.

Continental Shelf Act 1964, Chapter 29, HMSO, 1964.

Gas Council *Annual Report and Accounts*, HMSO (various years).

Kelf-Cohen, R., *Twenty Years of Nationalisation—The British Experience*, Macmillan, 1969.

Messham, Susan E., *Gas—An Energy Industry*, HMSO, 1976.

Petroleum and Submarine Pipe-lines Act, HMSO, 1975.

Report of the Committee of Enquiry into the Gas Industry, Cmd 6699 (Heyworth Report), HMSO, 1945.

The United States

Gas Facts, American Gas Association (various years).
Gas Rate Fundamentals, American Gas Association, 1960.
Economics of the Natural Gas Controversy—A Staff Study, Joint Economic Committee, Congress of the United States, 1977.
Federal Power Commission—*1977 Final Annual Report*, US Department of Energy, July 1978.
Natural Gas—A Study in Industry Pioneering, American Gas Association, 1963.
National Gas Survey, Volume I, Federal Power Commission, 1975.
Natural Gas Liquids, US Bureau of Mines, 1978.
Pipeline and Gas Journal, Petroleum Engineer Publishing Co., July 1976.
Story of Gas, American Gas Association, 1964.

Japan

Far Eastern Economic Review, various issues.
Gas Energy—its Present Situation and Future Course, Tokyo Gas Company Ltd, June 1977.
Gas Utility Industry in Japan, Japan Gas Association, various years.
History of the Gas Industry in Japan, Tokyo Gas Company Ltd, 1978.
Osaka Gas Company Ltd, *Annual Reports*, various years.
Tokyo Gas Company Ltd, *Annual Reports*, various years.

The Netherlands

Energy from the Depths, N.V. Nederlandse Aardolie Maatschappij, 1965.
Gas-Aardgas, N. V.Nederlandse Gasunie, 1972.
Gas Marketing Plan, N.V. Nederlandse Gasunie (various years).
Gasunie Annual Report (various years).
Kloosterman, A. H., *Development of the Natural Gas Pipeline System in the Netherlands*, International Gas Union, 1964.
Liquefied Natural Gas on the Maasvlakte, N.V. Nederlandse Gasunie, 1978.
Man and Energy, N.V. Nederlandse Gasunie, 1978.
Natural Gas, VEG-Gasinstituut, 1971.
Natural Gas in Holland, N.V. Nederlandse Gasunie, 1977.
Peebles, M. W. H., Pass, V. H., Salkeld, J., *The Development of Groningen Gas*, Shell International Gas Ltd, 1971.
Power from the Depths, Nederlandse Aardolie Maatschappij, B.V., 1973.
Voogd, J. G. de, *Natural Gas in the Netherlands*, The Institution of Gas Engineers, 1964.

The Union of Soviet Socialist Republics

Campbell, Robert W., *The Economics of Soviet Oil and Gas*, Resources for the Future Inc., 1968.
Ebel, Robert E., *Communist Trade in Oil and Gas*, Praeger Publishers, Inc., 1970.
Ebel, Robert E., *The Petroleum Industry of the Soviet Union*, American Petroleum Institute, June 1961.
Elliott, Iain F., *The Soviet Energy Balance*, Praeger Publishers Inc., 1974.
Hassman, Heinrich, *Oil in the Soviet Union*, Princeton University Press, 1953.
Oil and Natural Gas Industries of the USSR, Petroleum Economics Ltd, January 1959.
Orudzhev, S. A., 'The Development of the USSR Gas Industry', 13th World Gas Conference, 1976.
Park, Daniel, *Oil and Gas in Comecon Countries*, Kogan Page Ltd, 1979.
Russell, Jeremy, *Energy as a Factor in Soviet Foreign Policy*, The Royal Institute of International Affairs, 1976.
Soviet Oil Gas and Energy Databook, Petro Studies Co., 1978.
USSR: Development of the Gas Industry, Central Intelligence Agency, July 1978.

Liquefied Natural Gas

Coppack, C. P., *Practical Aspects of LNG Projects*, Australian Gas Association, October 1976.
History of the Methane Importation Scheme, The Gas Council and Conch Methane Services Ltd, October 1964.
Liquefied Natural Gas, Shell Briefing Service, July 1978.
Liquefied Natural Gas and the International Energy Market, Worldmark Economic Publications Inc., 1958.
Prew, L. R., *Technical Aspects of LNG and NGL Shipping*, Shell International Gas Ltd, 1977.
Perspective, Conch International Methane Ltd.

General

Adkins, J. M., *Town Gas from Petroleum*, The Shell Petroleum Company Ltd, 1958.
Annual Bulletin of Gas Statistics for Europe, United Nations, various years.
Asimov, Isaac, *Asimov's Biographical Encyclopedia of Science and Technology*, Doubleday & Co. Inc., 1964.
Carter, E. F., *Dictionary of Inventions and Discoveries*, Frederick Muller, 1966.

DeGolyer and MacNaughton, *Twentieth Century Petroleum Statistics*, 1976.

Lawrie, James, *Natural Gas and Methane Sources*, Chapman & Hall Ltd, 1961.

Natural Gas Terms and Measurements, Shell International Gas Ltd, 1969.

Pipeline and Gas Journal, Harcourt, Brace, Jovanovich Publications, November 1978.

Rose, J. W. and Cooper, J. R., 'Technical Data on Fuel', The British National Committee, World Energy Conference, 1977.

Short History of the German Gas Industry, Ruhrgas A. G., 1976.

Shnidman, Louis, *Gaseous Fuels—Properties, Behaviour and Utilisation*, American Gas Association, 1954.

World Energy Supplies 1971–1975, United Nations, 1977.

In addition to the foregoing, the following journals in particular, in their various issues over the years, contained a wealth of information on specific aspects of the gas industry which proved invaluable either in cross-checking or adding to that which was available elsewhere.

Gas World
Oil and Gas Journal
Oil Gas European Magazine
Petroleum Economist (formerly *Petroleum Press Service*)
Petroleum Intelligence Weekly
Petroleum Times
Platt's Oilgram News

Index